Submersible Sewage Pumping Systems Handbook

SWPA
Submersible Sewage Pumping Systems Handbook

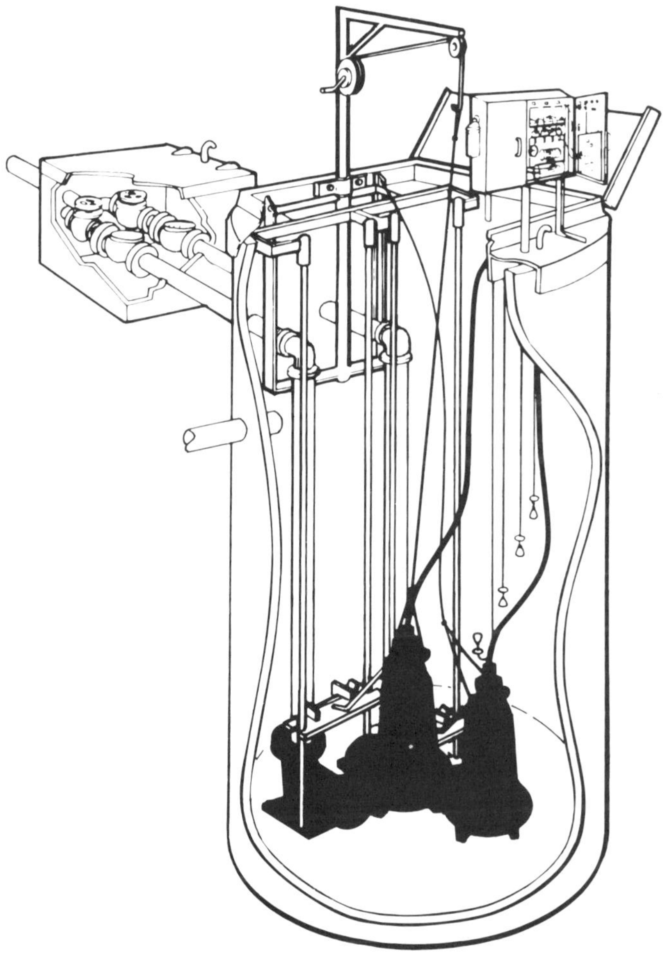

LEWIS PUBLISHERS, INC.

Library of Congress Cataloging in Publication Data

Submersible sewage pumping systems handbook.

   Includes index.
   1. Pumping stations—Handbooks, manuals, etc.
2. Submersible pumps—Handbooks, manuals, etc.
TD725.S83   1986   628'.29   86-20056
ISBN 0-87371-085-1

   1. Humes, Durward; 1928

This Handbook was developed by representative members of the Submersible Wastewater Pump Association to provide their opinion and guidance on the design and use of lift stations using Submersible Solids-Handling Pump Systems. This Handbook contains advisory information only and is a public service by SWPA. The reader is solely responsible for interpretation and use of this Handbook, and SWPA disclaims all liability of any kind for the use, application or adaptation of material published in this Handbook.

First hardbound edition, 1986

Published by Lewis Publishers, Inc.

    121 South Main Street
    P.O. Drawer 519
    Chelsea, Michigan 48118 USA

Second Printing 1987

Copyright© 1985
Submersible Wastewater Pump Association
600 South Federal Street
Chicago, IL 60605 USA

All Rights Reserved

Printed in the United States of America

TABLE OF CONTENTS

Introduction 1
1 Fundamentals and Components 3
2 Sizing the System 13
3 Selection of Submersible Pumps 25
4 Submersible Pump Controls 49
5 Mechanical Controls and Components 61
6 Installation and Start-Up 67
7 Operation and Maintenance 77
   System Glossary 85
   Electrical Glossary 89
   Appendixes 95
      Electrical Diagram Terminology 96
      Electrical Symbols 97
      Electrical Formulae 98
      Electrical Conversion Formulae 99
      Hazardous Location Basics 100
      Loading: Access Covers 101
      Metric Equivalents 102
      Friction of Water in Plastic Pipe 104
      Manufacturers 115
   Index 119

PREFACE

This Handbook evolved because the Submersible Wastewater Pump Association believed that too few specifiers, operators and managers of municipal sewage systems were sufficiently familiar with submersible solids-handling systems in lift station applications.

It was planned, drafted, written and re-written, and produced over a period of 30 months, in an intensive effort to make it both generic and immediately useful to its readers.

The principal contributing authors were:

A. Bo Andersson	Elmer Jira
Robert Brinley	David Maurer
Martin Dretel	David Runyon
John Drotar	Peter Worden
George Holmes	Bud Zocolo

Other contributing authors included Jim Bradshaw, Tim Jansen, Ken Nicholls, Ray Rhein, and other members of SWPA. The final copy was reviewed by Wayne N. Austin, Critical Systems Incorporated, Orlando, FL, and John W. Bottorf, Jr., P.E., Seabury-Bottorf Associates, Inc., Orlando, FL.

The Association thanks the companies who supported—with time, money and staff assistance—the principal contributing authors: The Bilco Company; Electric Specialty, Inc.; Enpo Pump Company; Flomatic Corporation; Goulds Pumps, Inc.; Hydromatic Pumps, The Marley Pump Company; Kenco Pumps, Inc.; Peabody Barnes, Inc.; Pumptron Division, Transamerica DeLaval, Inc.; and Putnam Water Guard, Inc.

Durward Humes of the Association staff furnished editorial assistance and supervised production.

Introduction

The purpose of the SWPA Handbook is to familiarize and assist those who are responsible for designing, installing and operating lift stations using submersible solids-handling pump systems.

It was prepared by a task force of knowledgeable specialists from submersible pump and accessory companies.

It presents fundamentals. It also addresses some of the more sophisticated aspects of submersible pump systems and their components...lift stations and selection of proper equipment...how to make certain the station is installed properly... and how to establish the operating and maintenance procedures needed to ensure long, satisfactory, and economic station life.

This Handbook is intended to be used as a daily reference. As a designer or specifier, it will give you the background on submersible pumping systems and — in Chapters 2 and 3 — provide detailed guidelines for sizing the station and selecting the proper pumping system. Chapters 4 and 5 discuss electrical and mechanical controls.

If you are responsible for installation, you'll want to concentrate on Chapter 6, which explains in detail how the entire station is to be put into place and made operational.

If yours is the responsibility for on-going operations, study Chapter 7, which offers specific guidelines on operation and maintenance procedures.

To make this Handbook more useful, there are pumping system and electrical glossaries. There are also a variety of source materials, including friction loss tables for various pipe and fitting configurations.

In summary, this Handbook presents the fundamentals and tells you how to use them in practice. If you are planning and designing a station, you will also need current manufacturer's literature relative to specific pump and component selection. If you are installing or maintaining a station, you will need specific equipment instructions from manufacturers.

We dedicate this Handbook to lift station designers and operators, and to broader use of reliable, efficient and economical submersible pumping systems.

Chapter 1
Fundamentals and Components

Submersible wet well lift stations are in wide use. They offer a number of advantages over wet well/dry well stations. There are Very Small, Small, Medium and Large Stations, each with their own design parameters. Special submersibles, called grinder pumps, are suitable for pressurized sewage system applications. Submersible system components include the station itself, the pump-motor units, piping, valving, electrical controls, and access frames and covers. Guide rail systems are unique to submersibles. Considerations in selecting a station site are discussed.

The purpose of a lift station is to receive, store temporarily as needed, and move wastewater and/or stormwater through a collection system.

The two generic types of lift station designs are *wet well* and *wet well/dry well*. The latter, which was first used in North America, uses a wet well (also called a pit or tank) to collect wastewater and a companion dry well which contains the pumping, control and related equipment.

The submersible wet well lift station, widely used in Europe, is now becoming more common in North America. It uses only one well as opposed to two. It is built around a submersible solids-handling pump-motor unit, which is designed to operate under water with an open submerged inlet. All system components are installed in or contiguous to the single wet well.

More and more, submersible solids-handling pumps are being specified in North America. They are used primarily for wet well sewage lift stations and for industrial sump or process effluent applications. A common application for smaller submersible pumps is to move effluent in septic tank systems. Large and small units are used in a variety of ways in home, farm, motel, school, marine, commercial building, industrial plant, and municipal sewage and stormwater systems.

Submersible pumps have proven effective over the last quarter of a century, refuting those skeptics who originally wondered how an electric motor-powered pump "could run under water." Originally developed in Europe, they are now used throughout the world to pump clear water, raw water, and wastewater. Millions are in daily use.

The submersible wastewater pump came to the U.S. about 1955. It became popular in the early 1960s, when a pump release system was developed to allow the pump to disconnect automatically from the piping and be lifted out of the well for service. This ended the dirty and sometimes dangerous task of sending men into the station to manually disconnect and then reconnect the pump.

There are two classes of submersibles. The first, commonly called *sewage ejectors*, are normally used in home and light commercial applications. They usually handle up to 2-in. spherical solids and range from 1/3 to 1-1/2 hp.

Larger submersibles, defined as handling greater than 2-in. solids, have a minimum of 3-in. discharge. They are used in municipal, commercial, and industrial applications, for pumping sewage and all types of industrial wastewater.

Fundamentals

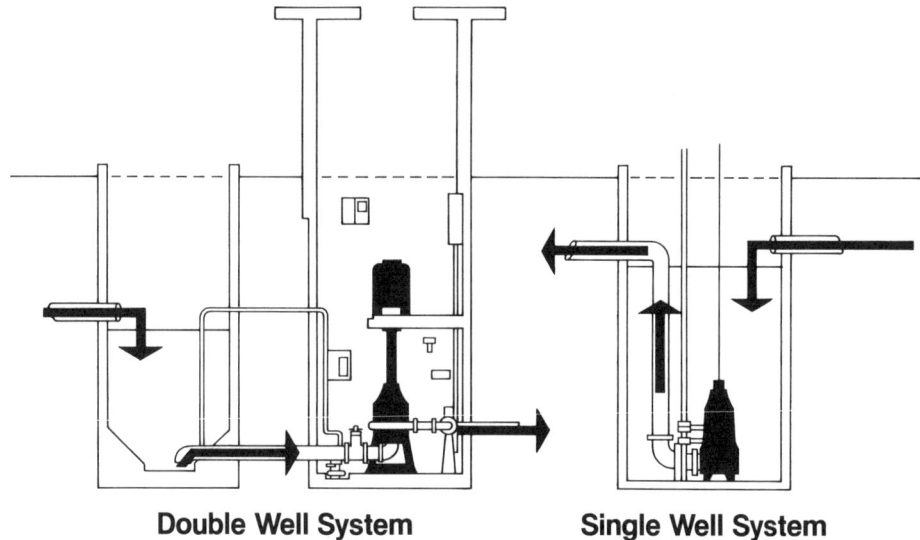

Double Well System   **Single Well System**

FIGURE 1-1: Wet well stations (right) take less space, use less equipment, and are generally less expensive to build than wet well/dry well stations.

Wet well lift stations, using submersibles, offer the following distinct advantages over wet well/dry well stations:

1) The single, buried well reduces the space needed for excavation and installation, and eliminates the need for elaborate above-ground buildings or enclosures.

There is seldom a need for dehumidifiers, lighting or other appurtenances which are normal for dry wells. The result is a substantial reduction in both construction and operating costs as explained in **Figure 1-1.**

2) Submersible pumps are specifically designed to operate in the liquid being pumped. This eliminates pump flooding problems.

3) Guide rail systems allow servicemen to remove, inspect and service the pump (or pumps) without having to enter the wet well. This reduces the chance of accidents during servicing.

4) Other servicing is easy, since all plumbing and electrical controls are accessible outside the wet well.

5) Controls can be installed directly over the station or in a remote location. There is a choice of control systems, from simple liquid level controls and alarms to sophisticated telemetry.

6) Submersible lift stations are designed to blend readily with natural surroundings, since there is no pump house and there's a minimum of above-ground equipment.

They are also less noisy, since the working installation is below ground level, and there is a reduced chance for the accidents which can occur with an exposed motor.

7) Above all, submersible solids-handling pumps are reliable and efficient to operate. Pumps run under and are cooled by the liquid, thus adding to their life span. They run only when needed, reducing wear and power bills. The system is designed to eliminate suction pipe clogging and net positive suction head problems.

Equipment and installation cost factors for wet well/dry well stations are summarized in **Table 1-2.**

Fundamentals

Station Classifications

Submersible wastewater pumping stations are usually classified by size, as follows:

Very Small Stations (less than 80 gpm). This type of station usually serves an individual dwelling or several isolated buildings with discharge lines smaller than 4-in. Simplex (single) pumps may be permitted, but duplex units are recommended for reliability and length of service. Both package and built-in-place stations are used.

Small Stations (80 to 300 gpm). Duplex stations are specified almost exclusively, with each pump sized to handle the maximum flow. Some locations allow 2½-in. solids but most states require a minimum of 3-in. solids-handling capacity and a 4-in. discharge. Provision is often made for future growth by selecting vendors who have the same pump models available over a range of impeller or horsepower sizes. This keeps expansion or conversion costs to a minimum on pump, control and sump changes. Alternately, provision is made for expansion with a third pump. Both prefabricated and built-in-place stations are used.

Medium Stations (300 to 3,000 gpm). Generally, duplex stations are used. However, more consideration is now given to expandable triplex systems which provide for future increases in pumping capacity. Each pump should be capable of handling more than the maximum design flow. Expandabil-

Comparative Needs And Costs

Equipment Costs

Equipment	Wet Well/Dry Well	Submersible
Pump	Equal	Equal
Electrical Controls	Equal	Equal
Discharge Piping	Equal	Equal
Suction Plumbing	Required	Not Required
Ventilation System	Required	Not Required
Sump Pump	Required	Not Required
Dehumidifier	Required	Not Required
Guide Rail System	Not Required	Required

Installation Costs

Description	Wet Well/Dry Well	Submersible
Wet Well	Equal	Equal
Dry Well	Required	Not Required
Excavation	Double	Single
Power Wiring	Equal	Equal
Accessory Wiring	Required	Not Required
Discharge Piping	Equal	Equal
Suction Piping	Required	Not Required

TABLE 1-2: Comparison of equipment and installation requirements for the two types of stations.

Fundamentals

ity is recommended, as noted under Small Stations. Discharge size diameter and line length become critical factors in providing for expansion. Prefabricated stations are still used in the smaller capacity units, but built-in-place stations dominate above 1,500 gpm.

Large Stations (over 3,000 gpm). Two pumps can be used, but efficiency of operation over varying ranges of flow usually dictates three or more pumps. Capacity should be selected so that, with the largest pump out of service, the others can handle the maximum flows.

As an example, resort areas which have seasonal variations in flow often require unequal horsepower combinations, since it is less expensive to run a 40-hp pump at 100 percent load than a 75-hp pump at 55 percent load. Rectangular stations with splash baffles on the interior walls to dampen water surge are necessary for high capacity pumps. Built-in-place, special design stations are commonly used.

System Components

Submersible solids-handling pump lift stations consist of:
- Pump-motor unit
- Station, also called the wet well or tank
- Piping
- Valve box, with mechanical controls
- Electrical panel and controls
- Pump removal apparatus, usually called the guide rail system
- Access frames and cover, both for the well and the valve box

The primary components of a wet station are shown in **Figure 1-3**. A standard station configuration is shown in **Figure 1-4**.

Most lift stations are designed for multiple pump installation. The most common is the duplex installation, where two units alternate in operation, to equalize the wear on the pumps. Multi-pump installations ensure continued operation if one pump fails or is removed for servicing, and also provide extra capacity in times of extraordinary loads.

Let's discuss each component of the installation.

Construction

   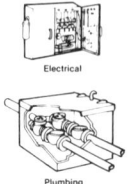

Pump Station Guide Rails Controls

FIGURE 1-3: Major components of a submersible lift station.

Pump-Motor Units

Submersible wastewater pumps are vertical, close-coupled, extra-heavy duty units which are designed to operate in the wastewater they are pumping. Since sewage contains solid material, they are designed to be non-clogging. They are commonly called *submersible sewage pumps*.

A special design of submersible, called the *grinder pump*, is frequently used for pressurized sewage systems. This pump incorporates a hardened stainless steel cutter designed to cut solids in the sewage into small pieces so that the slurry can be pumped under pressure through small diameter, low pressure force mains. Pressurized systems are easy to install. They normally use plastic piping, installed just below the frost line, and are especially adaptable to hilly or rocky terrains.

Grinder pumps were first used to serve single or multi-family dwellings with small (1, 1-1/2 or 2 hp) pumps where a septic tank or gravity design could not be used, or was too expensive. Larger horsepower grinders have now been developed to service small apartments, motels, nursing homes, restaurants, and similar commercial developments; they have the same advantages as small grinder systems.

Discharge Size: A standard submersible non-clog pump with a 4-in. discharge will normally handle 2-in. to 3-in. spherical solids. Each manufacturer's literature specifies the maximum solids size which can be handled by their particular pumps.

Discharge sizes for submersibles usually

Fundamentals

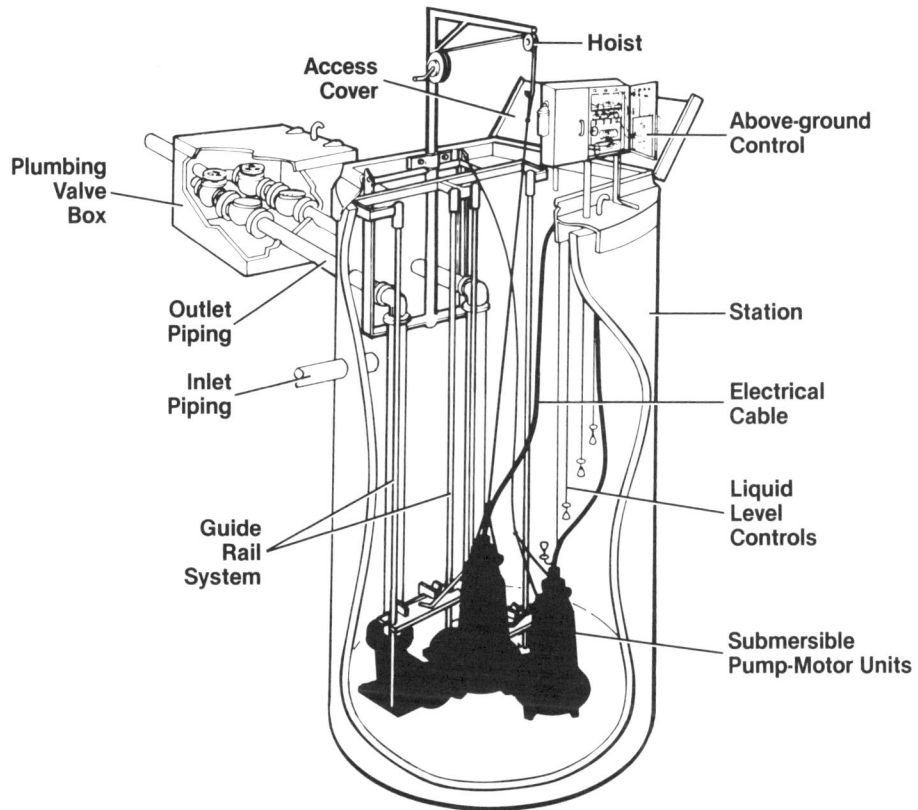

FIGURE 1-4: Configuration of submersible lift station, with component parts in place. Some stations are pre-fabricated.

range from 3-in. to 14-in., although larger sizes are available.

Motors: Submersible pumps must be sized to the application. Motors are normally available in 850, 1150, 1750, and 3450-rpm designs, using 60 cycle power. Horsepower ratings range from 1/2 to 100 hp and larger. Variable speed controls are also available with the use of variable frequency drives.

Depending on the application, motors generally operate on 208, 240, or 480 volts. Occasionally, 575 and higher voltage units are required. Motors may be single or three-phase; however, single-phase units are usually limited to 5 hp or less.

Capacities: Submersible pumps can be tailored to the capacity requirements of any particular installation. A specifier can ask for variable combinations of head and capacity as a function of the system head curve. Selection can accommodate such factors as existing and future flow requirements, discharge pipe size, materials, fixtures, and length of sewer line.

Typically, dynamic heads range from 15 to over 200 ft., and flows range from 50 to 7,000 gpm.

Submersible pumps are shown in a station in **Photo 1-5.**

Fundamentals

Wet Well Stations

Stations are available either as factory-assembled *(prefabricated)* packages, or are *built-in-place*. Both types can be a combination of standard components or can be custom designed.

Prefabricated stations are available from several manufacturers. They are usually made of coated steel, fiberglass or pre-cast concrete.

Station design should be based on both the current anticipated flow and projected future needs. The intake and discharge port levels will be dictated by the depth of setting of the station and the requirements of the sewage piping. The pump base plate, discharge piping and guide rail system are usually installed at the factory. Tank diameters are normally five to six feet, with a maximum of 10 feet to keep within practical over-the-road trucking limitations. The specifier should work closely with the pump manufacturer to determine final specifications for the prefabricated station.

Built-in-place stations are usually made of concrete. The same considerations apply — the final station must be designed and built to accommodate all wet pit system components.

The sizing of the station and pumping equipment is covered in detail in Chapter 2.

Piping and Controls

The intake port of the submersible station, as stated earlier, is located at the point where the incoming sewage line connects with the station. The liquid sewage flows to the bottom of the station. When the liquid level rises sufficiently, a control energizes the pump-motor unit, the sewage passes into the pump impeller, and is forced out through the discharge pipe and into the outgoing sewer line.

Thus, only one section of internal piping is required for each pump — from the pump discharge port to the station discharge port. To accommodate the removal and reinsertion of each pump for servicing, the pump discharge is sealed rather than bolted permanently to the discharge piping.

PHOTO 1-5: Submersible pumps in place and working in station.

The plumbing valve box is normally located adjacent to the station, at ground level, for easy access. There are check and shut-off valves for control purposes.

Electrical Controls

The heart of the control system for submersibles is the *liquid level control,* which turns the pump On when the liquid reaches a set level and turns the pump Off when the liquid drops back to a low level. Duplex pumps usually alternate operation automatically on each successive cycle. There is an override control which brings in the second pump when in-flow is unusually heavy or the first pump fails to operate.

Control panels are installed above ground, and usually contain: (1) pump disconnects, (2) across-the-line starters or contactors with overload protection, (3) hand-off-automatic selectors (H.O.A.), and (4) alarm systems to indicate high liquid level conditions in the wet well.

Alarm systems can be visual, audible or provide remote monitoring through telemetry devices or telephone lines.

The control manufacturer will help determine what controls are needed for a particular application, and then fabricate the control panel to this specification. All panels should be built in accordance with the National Electrical Code.

Guide Rail System

The pump removal and replacement apparatus, usually called the *guide rail (or*

Fundamentals

cable) system, is unique to submersible installations. It permits the pump to be removed, inspected and serviced, and then reinserted in operating position — with no need for the serviceman to enter the station.

The guide rail system is a standard component of the wet well installation. The rails, usually galvanized pipe, are mounted in place, from the bottom of the station to the access cover. A bracket on the pump unit mates with the rails. During initial installation, the pump or pumps are lowered into place on the railing and fitted to the discharge pipe by means of a *quick-disconnect sealing flange.*

A hoist is normally used to remove the pump for inspection or servicing. It may be portable, truck-mounted or be a permanent part of the installation. A chain or cable is attached to a bracket or to the top of the pump, the pump is lifted, and the sealing flange disengages from the discharge piping. The pump is then raised to the top of the station, guided by the rails. After inspection or servicing, the pump is lowered back into the wet well along the rails and the sealing flange automatically re-engages the discharge piping.

Access Covers

Access covers or doors complete the wet pit installation. They permit visual examination of the interior of the wet well, removal and reinsertion of the pump(s), and access to the valve box controls.

Access covers are locked in the closed position when not in use, but may also be locked in the open position to avoid having them close accidentally when the station is being serviced. They are available in steel and aluminum, as well as with special finishes for particular environments. They are also available in various sizes, styles, load-carrying capacities, and with lift springs for safe, easy manual operation.

Selecting a Site

Determining the best location for a pumping station (see **Photos 1-6 and 1-7**) begins with a comprehensive survey of the present and future needs of the area that the station will serve. The survey must consider how the location will affect construction costs, as well as the environmental impact of the station.

The survey should include a study of the topography of the land downstream from the proposed station, to determine if tunneling through high ground or providing interceptors could produce gravity drainage, and thus eliminate the need for the station. Other factors to consider include site conditions, land ownership, site and local drainage, traffic patterns, access for vehicles, and availability of utilities and community services such as electric power and telephone service.

If the survey confirms the need for the station, it may be best to locate it somewhere other than at the lowest surface elevation within the drainage area. For example, the survey may reveal that a remote location may be better than one in a densely populated area. Good site selection results from balancing technical needs, economic factors, and environmental impact.

The station must meet all local building ordinances and requirements. When specifying station size, consider the superstructure (if any), access and parking requirements, requirements for moving equipment in and out of the station, screenings disposal, property line clearances and the amount of excavation needed. Make certain that alarms and signals can be seen and heard.

The depth of the station substructure below the ground level is determined by the depth of incoming sewers or drainage chan-

PHOTO 1-6: Attractive lift station.

Fundamentals

PHOTO 1-7: Smaller lift station.

nels and by foundation conditions. Surface conditions, environment, and possible flood conditions determine the type of superstructure, height of electrical controls, and the degrees of exterior finish and trim. Because odors may leak from a pumping station, proper ventilation must be part of the design.

The critical wet well high level and bottom elevation are determined by the elevation of the basement of the residence most likely to flood and the elevation of the incoming sewer. Such a determination includes the hydraulic gradient and storage capacities of the wet well and sewer system.

Small lift stations are frequently built entirely underground. They can be prefabricated or built in place.

When stations with superstructures must be built in densely populated areas, it is important that the station be architecturally compatible with the other buildings in the neighborhood. Landscaping should complement the station's surroundings. In residential areas, tall and closely planted shrubbery can buffer noise and make the station less obtrusive.

In remote areas, the esthetic design is up to the discretion of the engineer. It must be consistent with erosion and sediment control practices used in the area. Fencing, burglar alarms, and the use of few or no windows in any buildings can provide protection against vandalism.

Site Considerations

Care must be taken by the installing contractor in setting and landscaping the lift station. The top of the station should be above grade at all times, and landscaping should insure that drainage water flows away from the station.

When the station is to be located in a flood-prone area, every effort should be made to relocate it. When this cannot be done, flood-tight or gas-tight access covers should be specified; while costly, they are available from specialty manufacturers. In this situation, the electrical controls must be mounted so that they will not be submerged at the highest known flood level.

In flood-prone locations, the electrical conduit entry must be completely sealed to prevent ground water from entering the junction box from outside the station. The preferred method is to locate the control panel on a telephone pole or pedestal above the flood plain, and to make sure there are no splices in the pump or level control cables.

Another site selection problem involves locating the lift station in streets, highways or other areas subject to traffic. Every possible alternative site should be considered. If this location is unavoidable, a manhole cover rather than a fabricated cover should be used.

Since station failure can cause considerable damage, it is important to specify dependable controls and an emergency power source for most wastewater and stormwater pumping stations.

The pumping station may be provided with a means of bypassing the wet well for maintenance or for emergency conditions. For these contingencies, a manhole can be constructed just upstream of the wet well, with provision made for a temporary pump. A valved connection is then provided in the force main just downstream of the pump station. A hose can then be used to link the

Fundamentals

pump to the valved connection in the force main, to direct the flow around the pump station.

During planning, long before construction begins, study the subsurface conditions of the station site. Analyze the profile and physical properties of the soil to determine the proper foundation. This will also define the chemical characteristics of the soil and the ground water conditions. Use this information on the ground water table to design drainage and/or dewatering systems for the station site, and to evaluate station design. Qualified geologists are best capable of analyzing and interpreting this data for foundation designers.

It is important to control the ground water levels during excavation. Inadequate or improper dewatering can lead to construction problems or failures. Dewatering begins with a detailed subsurface investigation to determine how to design the dewatering system. Consult a qualified excavating contractor before proceeding.

Chapter 2
Sizing The Station

Considerations in sizing a municipal sewerage system include a proper design period, projections of population growth and land development trends, and the quantity of available water and its consumption rate. Factors in sizing stations include average daily flow, maximum and minimum daily flows, and peak hourly flow. With residential stations, designers have the choice of grinder, STEP, or solids-handling systems. For larger systems, different sizing approaches are explained.

Sanitary wastewaters consist primarily of used portions of the public water supply that are discharged into sewers through the plumbing system of buildings. They sometimes include used portions of the private water supply from industrial and commercial establishments and residential communities. Before their ultimate disposal, these wastes normally are collected and carried to a plant for treatment.

To provide economical and effective treatment for these objectionable wastewaters, the system must keep the volume of sanitary waste to a minimum. Wastewaters such as storm runoff from roof, yard, and foundation drains should be separated from the sanitary wastewaters and then conveyed to discharge by a storm sewer system.

Design Considerations

Prior to the design and construction of any sanitary wastewater collection or transfer system, an engineering study is necessary. This study should analyze design periods, population growth projections, land development patterns, quantity of available water, and quantity of wastewater flows.

Design periods. There is a maximum hydraulic or functional capacity beyond which any wastewater system cannot operate satisfactorily. The time span — usually a number of years — from the date of the original design to an estimated future date when the maximum capacity of the system will be reached is referred to as a "design period."

The design period for any system should be determined before the start of the engineering design. This analysis should take into account obsolescence and the wear and tear on the equipment, the possibility of future expansion of the system, the ease or difficulty of carrying out such expansion, and the efficiency of system performance during its early years when it will not yet be operating at full capacity.

In choosing design periods, imposing an undue financial burden on the system's present users should be avoided. Careful selection of the design period should assure that system cost will be shared equitably by both existing and future consumers.

Population growth projection. The quantity of sanitary wastewater generated by a community depends on the population and the per capita wastewater contribution. Therefore, an accurate population projection is essential to a well-planned collection and treatment system. A community's population growth and distribution may be affected by employment opportunities, commuting distances for workers, socio-economic factors, zoning, availability of community facilities and services, etc.

Sizing

There are, however, other influences that may not be evident at the time of the study but that may emerge later to cause a major shift in population growth. Examples might be a new discovery of available natural resources in the vicinity or an industry's decision to locate in the community.

Details of population projection methods are discussed in many textbooks and references. Most of the graphical or mathematical projections are based on the extension of the past trend of growth. These methods are most applicable to communities of stable growth. For young and fast-growing communities, or communities with high growth potential, a meaningful projection requires good historical data coupled with a study of all related environmental determinants and the best possible engineering judgment.

In addition, population projections often can be obtained from previous studies. Earlier construction grant applications usually contain population projections for the community. Studies for larger basin areas will require further analysis to apply their results to a specific community. Environmental Protection Agency (EPA)-funded projects for wastewater treatment plant construction require that, for comparison purposes, federal population projections be adjusted to cover only the study area.

Land development. Local and regional land-use planning, development of urban renewal projects, and zoning regulations will influence the future population distribution. These in turn affect the quantity of wastewater to be removed from a community. Recent trends in construction of multi-dwelling apartments and townhouses have affected local densities and thus the quantity of wastewater produced.

Newer housing units often are equipped with modern appliances (dishwashers, garbage disposal units, clothes washers) that tend to increase the water consumption as well as the amount of sanitary wastewater generated; therefore, the age of housing units also must be considered.

Quantity of available water. The quantity of wastewater flow hinges mainly on the quantity of available water and its consumption rate. The general public's tendency to use water depends on whether the water supply is metered, the cost of water and availability of public sewers.

Because wastewater consists primarily of used water, the water consumption per capita should be established. Then the quantity of water reaching the sewer line can be estimated. Fluctuation in water demand by peak hours or even by season also plays a important role in design of sewerage systems and hence in the capacity requirements for pumping stations.

Quantity of wastewater flow. The quantity of flow varies continuously in a wastewater system. In any system where sanitary wastewater is the major contributor to the total flow, there will be a wide and perhaps an hourly fluctuation. If wastewater works are not designed to accommodate this, flow variation can cause operational problems in collection systems as well as in treatment facilities.

Approach To Sizing

Generally, in sizing a lift station, engineers study the average daily flow, maximum and minimum daily flows, and peak hourly flow. *Peak hourly flows* and *minimum daily flows* are used to evaluate the sewer system and conduits in treatment works. The *maximum daily flow* dictates the size of flow equalization facilities, and the sizing of wet wells and pumping equipment. The *average daily flow* helps to analyze the performance of a plant's treatment units. (Special sizing considerations may be required and will be discussed later in this chapter.)

To make calculations using these design criteria to size a sewage lift station for a city, municipality or town, information is needed on the number of persons being served and the water consumption per capita per 24-hour day.

This flow chart (**Figure 2-1**) is typical of the flow from a given town or city, and will be used *simply as an example for our calculations*. Actually, the change in flow during a 24-hour period — the minimum and max-

Sizing

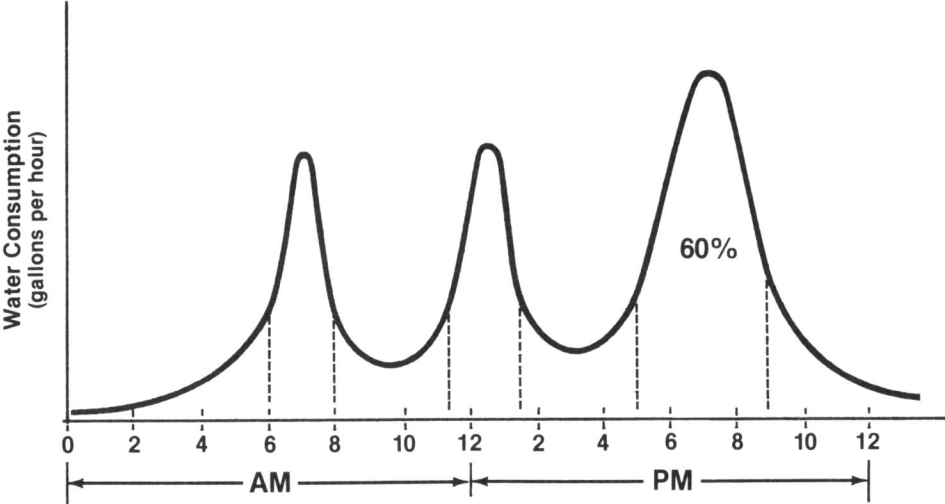

Figure 2-1: Sewage flow pattern in a typical city or town.

imum and their duration — will differ depending on location and type of environment. The important point is that if the city or town can present a reasonably accurate flow chart, a submersible lift station can be designed to do the job in an efficient and economical manner.

This particular diagram shows that during the 24-hour period, the waste consumption has three peaks:

1. Morning 6:00 a.m. — 8:00 a.m.
2. Noon   11:30 a.m. — 1:30 p.m.
3. Evening 5:00 p.m. — 9:00 p.m.

The evening peak is the largest and shows that approximately 60 percent of the total water consumption is concentrated in this four-hour period.

The station should be sized according to the inflow during this peak period:

$$Q_{in} = \frac{60\% \text{ of 24-hr. flow in gallons}}{\text{the 4-hr. period in minutes}} = \text{gpm}$$

$$Q_{in} = \frac{.60 \times n \times f}{4 \times 60} = \text{gpm}$$

Where: Q_{in} = inflow to the station in gpm

n = number of persons being served

f = water consumption in gallons per capita per 24-hour day

.60 = 60% as shown in **Figure 2-1.**

For design purposes, it is necessary to use existing conditions to estimate future wastewater flows. If the system has no flow records or if a complex sewer network makes measurement impractical, design flows may be determined by estimating and totaling flows from the various contributors. Or the flow may be calculated as a percentage of the total available water supply.

Careful study is needed, however, to project wastewater flow. Average flows may vary from 50 to 210 gallons per capita per day (gpcd). Weather conditions, water use habits, access to public sewers, water and sewer cost rates, adequacy of water supply, and the economic condition of the community all contribute to this wide variation in per capita flow.

Sizing

In addition to domestic wastewater flows, contributions may come from commercial, industrial, institutional, and infiltration/inflow sources.

Residential Stations

The residential lift station can use either a simplex or duplex grinder, Septic Tank Effluent Pump (STEP), or solids-handling pump system. All three can pump into gravity sewers or pressurized systems.

There are a number of advantages to pressure sewers. Because they use small-diameter plastic pipe buried just below the frost penetration depth, their installation cost in low-density areas can be quite low compared to conventional gravity systems. Other site conditions that enhance this cost differential include hilly terrain, rock outcropping, and high water tables.

Because pressure sewers are sealed conduits, there should be no opportunity for infiltration. Treatment plants can then be designed to handle only the domestic sewage generated in the homes serviced, which reduces the cost of processing the infiltration that occurs in most gravity systems.

As with any technology, there are also certain disadvantages. For pressure sewers, these include added operation and maintenance costs because of the need for mechanical equipment at each point of entry to the system. In addition, the wastewater conveyed through the treatment facility may be more concentrated than normal — and may require a higher level of treatment to satisfy effluent requirements.

Essentially, a pressure sewer system is the reverse of a water distribution system. Where the latter employs a single inlet pressurization point and a number of user outlets, the pressure sewer embodies a number of pressurized inlet points and a single outlet. The input to the pressure main follows a generally direct route to a treatment facility or a gravity sewer, depending on the application; the primary purpose of this type of design is to minimize the time that sewage is retained in the sewer.

Illustrated are the grinder or solids-handling pump system (**Figure 2-2**) and the STEP system (**Figure 2-3**).

Tank diameters for simplex grinder and solids-handling pumps are generally limited

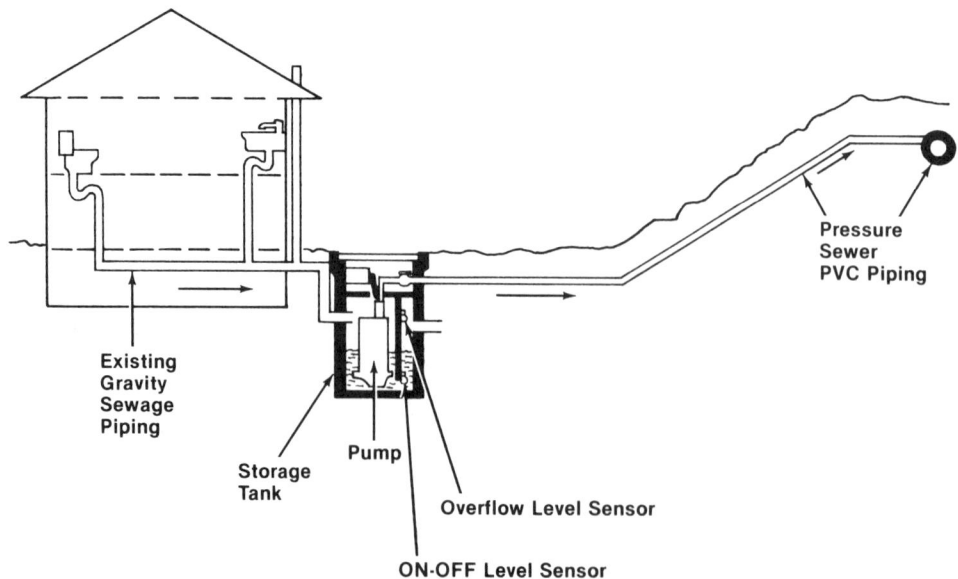

Figure 2-2:   Grinder/solids-handling pump system.

Sizing

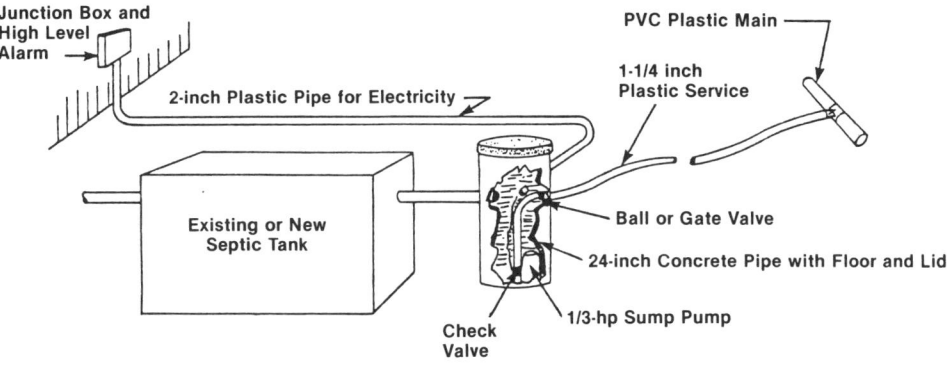

Figure 2-3: Septic tank effluent pump (STEP) system.

to 24 to 30 inches, while depths are regulated by soil conditions and/or site elevations. The general practice is to provide one tank per home, but sometimes two and even three homes will be connected to a single tank. The problem is not so much whether the system can handle the additional load as the logistics of connecting piping from several homes.

The grinder unit reserve capacity is shown in **Figure 2-4** as the volume available for storage between the On level float switch and the in-flow. Several states have established standards for reserve storage capacity; with a simplex grinder pump, this volume is usually 50 gallons.

The STEP system features a fractional horsepower sewage/effluent pump operating in conjunction with a septic tank. There are various designs, ranging from use of the existing septic tank with a new effluent tank **(Figure 2-5)** to a single tank system in either concrete or fiberglass **(Figure 2-6)**. Septic tanks generally have about 100 to 200 gallons of residual capacity due to the free board (septic storage) inherent in their construction.

Small, solids-handling pumps for residential use are normally fractional to two horsepower in size. They normally use single-phase, 115 or 230-volt motors, and are designed to pass up to 2-in. solids. Like the STEP pump, they operate at 1,750 rpm and can be made of cast iron, bronze and/or plastic. Unlike effluent pumps, they have a solids-handling impeller and volute.

Medium Stations

Very small, small and medium-sized submersible lift stations incorporate one to three solids-handling sewage pumps with capacities up to 3,000 gpm. The overwhelming majority of applications use two

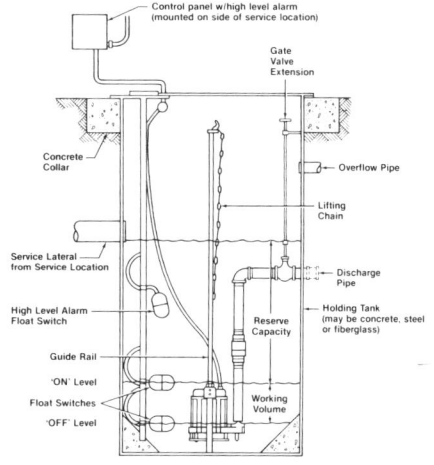

Figure 2-4: Grinder/solids-handling pump station.

Sizing

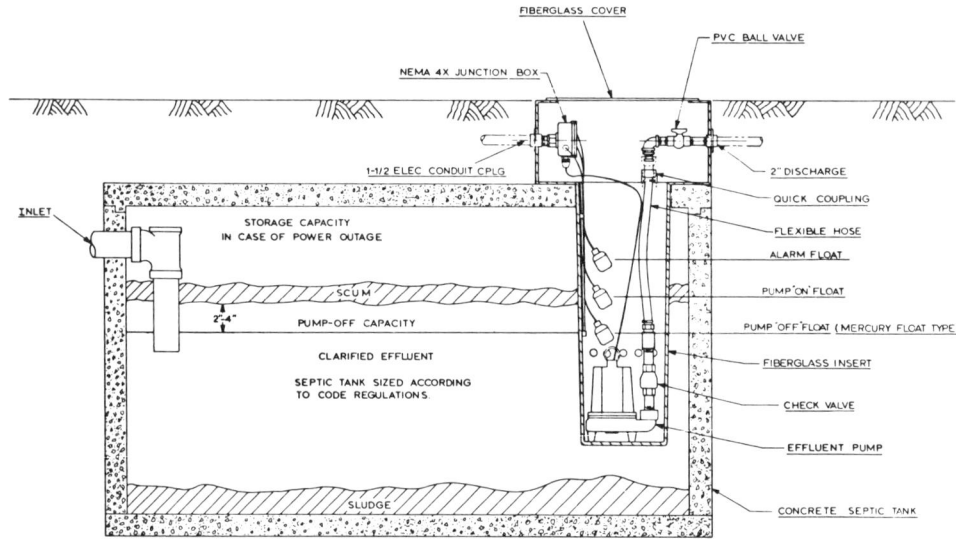

Figure 2-5: Two-tank STEP station, using existing tank and new effluent tank.

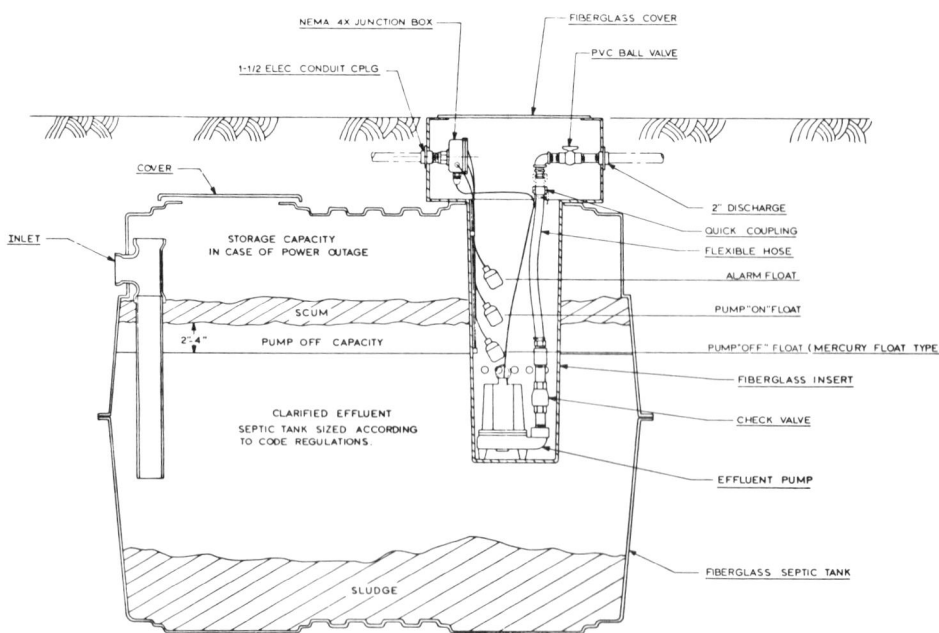

Figure 2-6: Single-tank STEP station, in fiberglass or concrete.

Sizing

pumps, with one pump sized for peak flow conditions and the second pump designed for 100 percent standby. Control panel design alternates pump operation so both get equal wear.

Minimum submersible duplex station dimensions are determined by the physical size of the pumps, the discharge piping, and the controls. Total station volume — a function of peak flow and pump capacity — is the paramount factor.

Critical considerations to be used in sizing the submersible lift station are as follows, referring to **Figure 2-7**:

Plan Area A: This area is established by calculation, but may be influenced by the dimensional requirements for the pumps, piping and controls.

Height X: This dimension is normally specified by the pump manufacturer and is generally defined by the top of the volute. It is designated as the Stop or Shut-off level of the pump. The purpose is to preclude drawing air into the pump inlet and/or prevent vortexing. The volume resulting from this height is a residual volume, and cannot be included in the working or storage volume of the station.

Height H_{min}: This is the distance from the pump Start level to Stop level. When multiplied by Plan Area A, it defines the volume which will be removed from the station during each pump operating cycle. Note that the volume occupied by the pump and other equipment located in this area must be subtracted to obtain the usable or true volume of fluid removed. The net result defines the minimum recommended pump cycle volume.

Height H_{lag}: The vertical distance between the turn On of the lead pump and turn On of the lag pump. This provides system protection should the lead pump fail or be incapable of keeping up with the inflow.

Height$_{res}$: This is the height measured from the Lag level to the level of the inlet pipe. This height can be used to determine the reserve capacity available in the

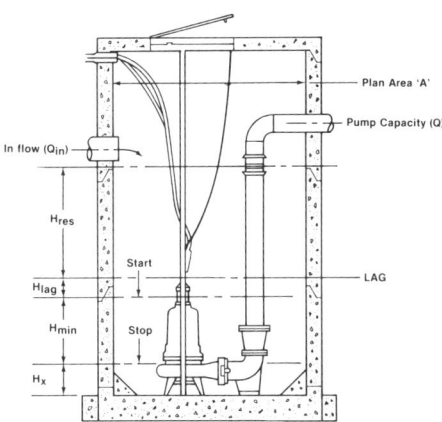

Figure 2-7: Submersible lift station design considerations.

station — a capacity available for emergency conditions, such as a power outage. Note that the reserve capacity is restricted to the station only and should not include inlet lines.

The consulting engineer may choose a reserve volume that differs from the above. For example, the minimum pump cycle volume can be increased as a safety factor. This volume should not be decreased by any significant amount, however, as there is a danger that the cycle time (pump On/pump Off) may be too short, which could shorten normal pump life.

To keep the size of the lift station at a reasonable minimum, the pump capacity should be two times the inflow at critical or peak flow times. In other words:

$$Q = 2 \times Q_{in}$$

Where:

Q = capacity of pump in gpm

Q_{in} = peak inflow to the station in gpm

Normally the station would be a duplex configuration, with a second pump that will alternate with the first unit, to assist at extreme peak flows and provide a standby in case of failure of one pump.

When sizing the pumps, the velocity in the piping system must also be considered.

Sizing

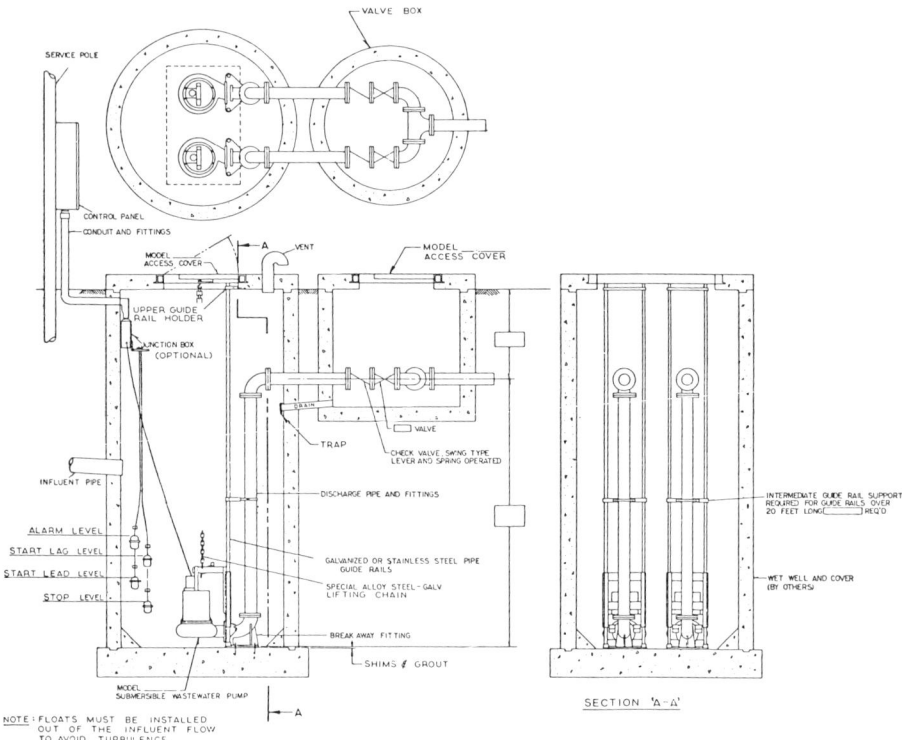

Figure 2-8: Typical duplex submersible station. (Courtesy: Peabody Barnes, Inc.)

Generally, two to three feet per second (ft./sec.) is considered sufficient to prevent sedimentation. This value can have a definite effect on sizing the pump. For instance, for 6-in. pipe, there should be a minimum flow of approximately 200 gpm.

The preceding pump calculations are fairly simple and will, in general, result in proper sizing of small to medium lift stations. However, many consultants use different and more complicated methods involving leakage in pipe systems, rain or storm volumes, and future growth requiring increased station and pump capacity.

Pump Cycle Volumes

The following calculation will give pump cycle volumes for a typical lift station. It can be used for stations with one to three pumps and a total pumping capacity of up to 3,000 gpm.

Use the formula: $V_{min.} = \dfrac{T_{min.} \times Q}{4}$

Where:

$V_{min.}$ = the minimum effective pump cycle volume in gallons. This is the volume between the Start and Stop levels in the station. The Stop level is normally at the top of the volute. The Start level should be somewhere below the inflow pipe to prevent possible sedimentation in the inflow piping system.

$T_{min.}$ = the minimum cycle time in minutes, defined as the amount of time it takes to raise the liquid

Sizing

to the Start level and then to draw back down to the Stop level. This minimum cycle time is achieved when the pump capacity (Q) is two times the inflow (Q_{in}).

Q = pump capacity in gpm

Example:

Pump capacity (Q) is 500 gpm

Minimum cycle time is six minutes (based on 10 starts per hour, as recommended by most submersible motor manufacturers)

What is the minimum pump cycle volume?

$$V_{min.} = \frac{T_{min.} \times Q}{4} = \frac{6 \times 500}{4} = \frac{3000}{4} = 750 \text{ gals.}$$

The minimum cubic area of any pump station shall take into consideration the physical size of the pumps and piping, and the liquid they displace. The space between the pumps is generally a function of the piping, pump dimensions, and pump capacity; this is a necessary consideration to avoid restricting or obstructing inlet flow to each pump. Minimum spacing between pumps and minimum submergence, to prevent vortex conditions, must also be considered. (A typical duplex installation is shown in **Figure 2-8**.)

Large Stations

When designing large submersible pump stations — over 3,000 gpm — there are a number of additional considerations. When three or more high-volume pumps are operating in a station, care must be taken in positioning the pumps, to provide a clear flow to the volutes to avoid sedimentation, and in sequencing the operation of the pumps.

There are two generally accepted methods of performing the pump sequence. In operating sequence I **(Figure 2-9)**, the pumps in a station start in sequence, one after the other, and stop in the reverse order. In this case, a more uniform output is

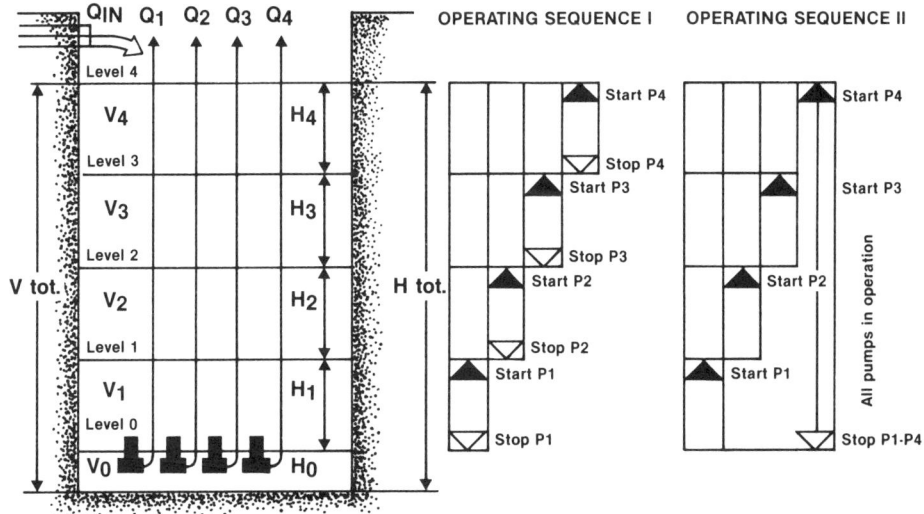

Figure 2-9: Two possible operating sequences in a multi-pump station. (Courtesy: Flygt Corp.)

Sizing

obtained from the pumping station than is the case for operating sequence II.

In sequence II, the pumps also start successively, but all of them then continue to operate down to the minimum Off level. *Operational sequence II always requires smaller pump sump volumes than sequence I.*

The starting order of the pumps should be alternated cyclically to distribute the running time of each as equally as possible.

A properly designed pump chamber ensures the even flow of water, without vortices or eddies, to the pumps. The inflow is distributed through the holes in the bottom of the inlet chamber opposite each pump.

Figure 2-10 shows a plan view of a four-pump station. **Figure 2-11** shows a side view of the same station. In station design, the primary considerations are the distance from the inlet to the center line of the pump(s), the distance between the center lines of pumps in stations with more than one pump, the maximum distance from the center line of each pump to the nearest wall, and the distance from the edge of the tank or station wall to the pump baffle.

When designing a particular installation, consult with the pump manufacturer. They

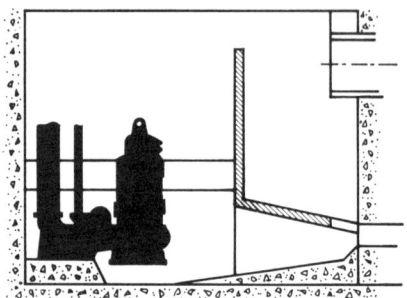

Figure 2-11: Vertical section in a station with one or more pumps.

can help prepare a layout which will insure the even flow of water to the particular type and style of pump used.

Any air bubbles which enter with the water into the pump chamber rise upwards along the sloping underside of the inlet settling chamber and escape from the surface of the water near the vertical partition wall.

Since all of the water is in motion, there is little risk of sedimentation so long as the minimum dimensions are not exceeded by any significant amount.

To obtain a larger sump volume, the best dimension to increase is the distance from the inlet chamber to the pumps. Since the water flows to the pumps over this path, sedimentation is prevented.

The minimum water level in the pump chamber — the minimum Off level for the pumps — must be high enough so that the square holes in the bottom of the inlet chamber are always submerged. In addition, it should be noted that the lowest water level is determined by the required Net Positive Suction Head (NPSH) for the pump and, in any event, should not be lower than the top of the pump casing.

The size of any particular station is determined by the number and physical size of the pumps, and by the output capacity per pump. When sizing a station, consult with the pump manufacturer. He can furnish specific recommendations for minimum clearances between pump casings and between the side wall and the pump casings.

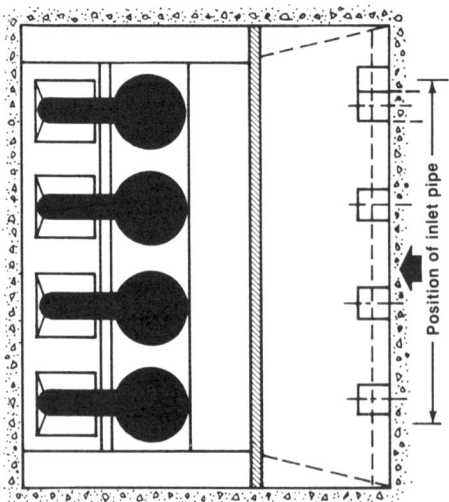

Figure 2-10: Horizontal section in a four-pump station.

Sizing

He can also recommend pump location relative to discharge connection base and the preferred distance between the pump sump and the pump inlet.

Estimating Flows

If actual system flows are available, they should be used as the criteria for sizing pumps and the volume of the sump. The preceding method is recommended.

However, the state, city or county may have a set of standards that consulting engineers must use. Wherever such standards exist — and they can differ depending on the area of the country or even in different parts of a state — they generally override any other method of computing station design. Although some of these standards are not as accurate as working with flow charts, they make it easier for the engineer.

When selecting the pump, some published standards simply say that in a two-pump station, each pump must be capable of handling flows in excess of the expected maximum flow.

A general standard for estimating flows is as follows:

Number of homes to be serviced, times
Average of 3.5 people per home, times
Average or 100 gallons per person per day.

Therefore, if a station were being designed for 200 homes, the 24-hour inflow would be:

200 × 3.5 × 100 = 70,000 gallons per day (gpd).

The average inflow to the station in gpm would then be:

70,000 gals. ÷ 1,440 mins. or 48.6 gpm.

Since both the station and the pump capacity must be sized as a function of the expected peak flow at any given time, the standard will call out a Peak Factor. The Average is then multiplied by the given Peak Factor. In the above example, with a Factor of 3.5, which is often used, the estimated peak flow would be 48.6 × 3.5 = 170 gpm. This appears to be simple but workable.

Chapter 3
Selection of Submersible Pumps

Submersible centrifugal pumps come in different designs, each with specific characteristics and capabilities to meet various operating conditions. Pump capacity is determined primarily by the speed, size and design of its impeller, which creates liquid head and flow through its rotating motion. Other factors in capacity are friction, leakage and shock losses. The volute — or stationary part of the pump — guides the liquid being pumped through the discharge opening. The size of the impeller passages and the clearance between the impeller and the volute allows for the passage of solid particles in the liquid. This chapter discusses the many factors in the selection of submersible wastewater pumps for particular applications.

A submersible centrifugal pump is a rotary machine consisting of two basic parts — the rotary element or impeller, and the stationary element or volute.

General Theory

A pump converts the energy provided by its motor to a combination of velocity and pressure energy through the impeller. The liquid being pumped surrounds the impeller. As the impeller turns, it imparts a rotating motion to the liquid.

The head developed by this action is a function of the difference between the impeller vane diameter at entrance and the impeller vane diameter at exit. The expression of theoretical head (at zero flow) can be related to the law of falling bodies:

$$H = \frac{V^2}{2g}$$

Where: H = height or head, ft.
V = velocity of moving body, ft/sec
g = acceleration of gravity (32.2 ft/sec²)

When the height of fall is known (for example, H = 100 ft.), the terminal velocity can be determined — in this case, V = 80.3 ft/sec. Conversely, if the direction of motion is reversed, a liquid exiting through an impeller vane tip at a velocity of 80.3 ft/sec reaches a velocity of zero ft/sec at 100 ft. above the impeller tip. (**Figure 3-1**)

When the developed head required is known, the theoretical impeller diameter for any pump at any rotational speed and zero flow can be determined from the following equation for peripheral velocity of a round rotating body:

$$D = \frac{(229.2)V}{N}$$

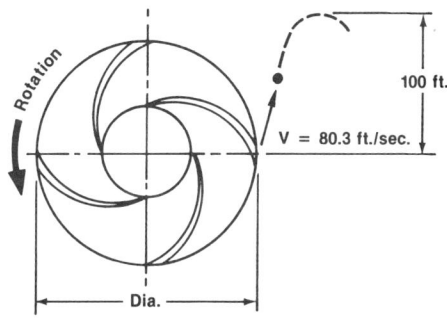

Figure 3-1: Law of theoretical head.

Selection

Where: D = unknown impeller diameter, in.
V = velocity (derived from $V = \sqrt{2gH}$), ft/sec
N = rotational speed of the pump, rpm

If the pump head is 100 ft. and the rotational speed is 1750 rpm, the formula indicates that a 10.52 in. theoretical diameter impeller is required.

In the preceding formula, we saw that if the speed of a centrifugal pump is doubled, the head developed by the pump is quadrupled. This is because the head developed is proportional to the square of the velocity change.

Doubling the operating speed of a centrifugal pump also doubles the capacity which the pump can handle. This is because the velocity of the fluid through the impeller has doubled. For example, a pump capable of developing 50 ft. total head at 100 gpm when operating at 1750 rpm with a given impeller diameter will be capable of developing 200 ft. total head at a capacity of 200 gpm when operating at 3500 rpm.

Let's go back to the formula relating head, capacity, efficiency, and brake horsepower:

$$\text{Head } H = \frac{V^2}{2g}$$

$$\text{Capacity } Q = VA$$

$$BHP = \frac{QH}{K \times \text{efficiency}}$$

Where: Q = capacity, ft³/sec
K = constant = 3960
A = cross sectional area, ft²

Assuming that the efficiency of the pump remains the same when we change the speed, we see that the horsepower increases by a factor of eight when the speed of the pump is doubled. This is because the capacity doubles when we double the speed, and the head quadruples — and these two factors are multiplied together to arrive at the brake horsepower.

For this reason, the speed of a centrifugal pump cannot be increased arbitrarily. A pump designed to operate at one speed must be capable of transmitting a great deal of additional horsepower if it is to run at a higher speed. However, we can make use of the speed conversion in slowing a pump down, from 3500 to 1750 or from 1750 to 1150 rpm.

To get a rough idea of a pump's capability, we assume the efficiency remains constant and apply the relationship given below, commonly called the *Pump Affinity Laws*.

For a given pump and impeller diameter:

$$\frac{rpm_1}{rpm_2} = \frac{gpm_1}{gpm_2} = \left(\frac{head_1}{head_2}\right)^{1/2} = \left(\frac{bhp_1}{bhp_2}\right)^{1/3}$$

Or, to put it a different way:

$$\frac{rpm_1}{rpm_2} = \frac{gpm_1}{gpm_2} \quad \text{or}$$

$$\left(\frac{rpm_1}{rpm_2}\right)^2 = \frac{head_1}{head_2} \quad \text{or}$$

$$\left(\frac{rpm_1}{rpm_2}\right)^3 = \frac{bhp_1}{bhp_2}$$

A change in speed always results in a change in capacity, head, and horsepower. These formulae give a useful approximation of the final performance capability of the pump.

A change in impeller diameter of a pump being operated at constant speed has essentially the same effect as a change in pump speed — both result in a change in the speed of the liquid which leaves the impeller. The formulae which apply for changes in capacity, head and horsepower as related to impeller diameter changes look exactly the same as they do for changes in speed.

Selection

For a given pump and speed:

$$\frac{\text{Imp. dia.}_1}{\text{Imp. dia.}_2} = \frac{\text{gpm}_1}{\text{gpm}_2} = \left(\frac{\text{head}_1}{\text{head}_2}\right)^{1/2} = \left(\frac{\text{bhp}_1}{\text{bhp}_2}\right)^{1/3}$$

Or, to put it a different way:

$$\frac{\text{Imp. dia.}_1}{\text{Imp. dia.}_2} = \frac{\text{gpm}_1}{\text{gpm}_2} \text{ or } \left(\frac{\text{Dia.}_1}{\text{Dia.}_2}\right)^2 = \frac{\text{head}_1}{\text{head}_2} \text{ or } \left(\frac{\text{Dia.}_1}{\text{Dia.}_2}\right)^3 = \frac{\text{bhp}_1}{\text{bhp}_2}$$

Even greater care must be taken in the use of these formulae than the formulae for change in speed. A change in the impeller diameter in a pump affects the basic relationship between impeller and volute, and this alters the design configuration of the pump. For this reason, the impeller change formulae should not be applied when the impeller diameter changes are more than about 10 percent. When larger diameter change is indicated, the best practice is to obtain new test curves from the manufacturer.

Specific speed is a correlation of pump flow, head, and speed at optimum efficiency. It classifies pump impellers with respect to their geometric similarity. Specific speed is expressed as:

$$N_S = \frac{N\,(Q)^{1/2}}{(H)^{3/4}}$$

Where: N_S = pump specific speed
Q = flow at optimum efficiency, gpm

The specific speed of an impeller is defined as the revolutions per minute at which a geometrically similar impeller would run if it were of a size that would discharge one gpm against a head of one ft. Specific speed is indicative of the shape and characteristics of the impeller.

Centrifugal pumps are divided into three classes — radial flow; mixed flow; and axial flow.

There is a continuum design change from the radial flow impeller (which develops head principally by the action of centrifugal

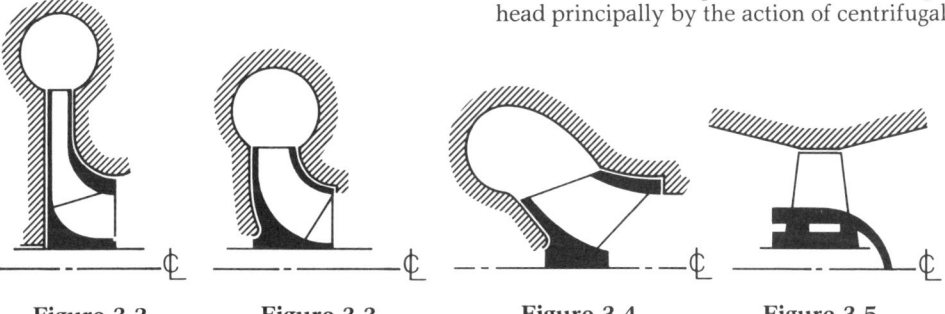

Figure 3-2   Figure 3-3   Figure 3-4   Figure 3-5

Figure 3-2: Radial type impeller. Normally for low capacities with medium and high heads. Specific speed range 500 to 3,000 rpm. Figure 3-3: Francis type impeller. This axial inlet/radial discharge style is normally used for lower heads. Specific speed range 1,500 to 4,500 rpm. Figure 3-4: Mixed flow impeller with combination axial and radial inlet and discharge for low head/high flow. 4,500 to 8,000 rpm specific speed. Figure 3-5: Propeller-type impeller. For low head, low speed, and large volume. Specific speed above 8,000 rpm.

Selection

force) to the axial flow impeller (which develops most of its head by the propelling or lifting action of the vanes on the liquid).

Typically, centrifugal pumps can also be categorized by physical characteristics relating to the specific speed range of the design, as shown in **Figures 3-2 through 3-5.** Once the values for head and capacity become established for a given application, the specific speed range of the pump can be determined and specified to assure optimal operating efficiency.

Solids-Handling Design

When used in sewage handling installations, a further criteria is required in impeller selection — the ability to pass solid particles through the impeller passages. There is a wide choice of non-clog impellers in either semi-open or enclosed configurations, and either single, two or three-vane types. Vortex or recessed impeller designs are also available. An explanation of these types follows.

Semi-Open Non-Clog: An impeller with a solid back plate, to which the vanes are attached, is called a semi-open impeller. **(Photo 3-6)** Clearances between the impeller and volute can be restored by reshimming. A disadvantage is the axial load resulting from pressure imbalance between the open and closed sides; however, the bearings are designed to handle this load. **(Figure 3-7)**

Enclosed Non-Clog: When a solid plate is put on both sides of the impeller, a closed or enclosed configuration exists. **(Photo 3-8)** The advantage of this configuration is that a volute wearing ring is used, which results in higher pump efficiencies. The disadvantages are more complicated fabrication and the need for slightly larger axial space. **(Figure 3-9)**

Vortex—Multi-Vane: Non-clog operation can also be accomplished with a vortex or recessed impeller. With this configuration, a vane with a shortened height is used so that the vast majority of liquid passes between the top of the vane and the volute by whirlpool action, thus ensuring a definite non-clog condition. **(Photo 3-10)** The primary advantages of this impeller style are the reduced radial loading and the ability to pass long, stringy particles. The disadvantage is reduced efficiency. **(Figure 3-11)**

Volute

It is the function of the pump volute, often referred to as the *pump casing,* to collect liquid discharge from the impeller and direct it through the discharge nozzle or opening of the pump. The volute is designed so that, at one point, its wall is very close to the outer diameter of the impeller. This point is called the *cutwater* or *tongue* of the volute.

Between the cutwater and a point slightly beyond, a certain amount of liquid has been discharged from the impeller. This liquid

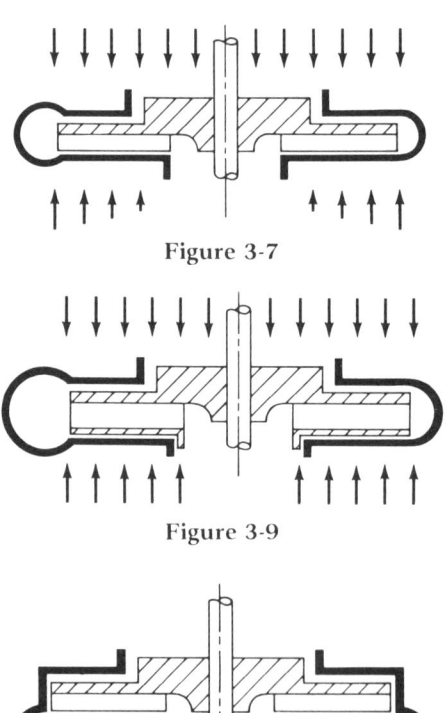

Figure 3-7

Figure 3-9

Figure 3-11

Figure 3-7: Graphic portrayal of unbalanced axial forces in a semi-open impeller. Figure 3-9: Unbalanced axial forces in an enclosed impeller. Figure 3-11: Vortex or recessed impeller.

Selection

Photo 3-6 (left): The semi-open impeller is easily fabricated. Its vanes are attached to a solid back plate. Photo 3-8: The closed impeller provides more efficient operation by eliminating spillover between passages. Photo 3-10: The vortex or recessed impeller prevents clogging by passing most of the liquid between the top of the vane and the volute by whirlpool action.

must rotate with the impeller until it is finally discharged through the outlet nozzle of the pump. Additional liquid is discharged from the impeller at every point around the volute, and this must also travel with the impeller and be discharged through the outlet nozzle.

As the liquid continues around the volute, more and more volume accumulates, which must be carried around between the wall of the volute and the outer edge of the impeller. To keep the velocity fairly constant, even though the volume of liquid increases, the area between the tip of the impeller and the volute wall is gradually increased from the cutwater to the beginning of the discharge nozzle. (**Figure 3-12**)

In a sewage handling (or non-clog) pump design, clearance must be provided between the impeller O.D. and the volute cutwater to permit passage of solid particles as they exit the impeller passage at the moment it aligns with the cutwater.

Pump Performance

Submersible centrifugal pumps are designed to operate over a wide range of discharge head conditions. A typical performance curve is shown in **Figure 3-13**.

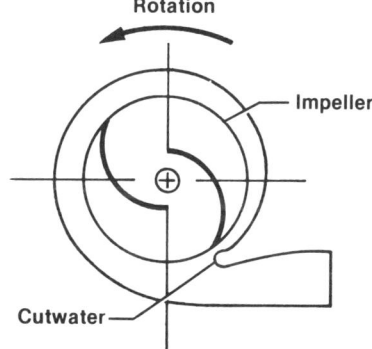

Figure 3-12: Volute and impeller relationship in a centrifugal pump.

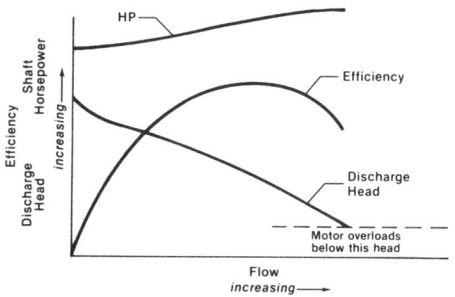

Figure 3-13: Typical performance of a centrifugal pump.

Selection

As this illustration shows, output flow and input horsepower increase as the discharge head decreases. The selection of a centrifugal pump for a particular application is usually made at the *best efficiency point (BEP)*, which normally falls in the midsection of the pump performance curve. Operation at either extreme of the curve should be avoided. If a pump is operated below the manufacturer's specified minimum head (high flow end of curve), motor overloading can occur. Conversely, pump selection in the extreme high head region of the curve results in low fluid flows and efficiency.

When sizing a centrifugal sewage pump, one has to consider pumping solids as well as the water. Solids require a minimum flow/velocity to keep them moving with the water in the pipe. Experience and test data have established this flow/velocity at 2 ft/sec.

Water, therefore, has to be pumped at a minimum flow of 80 gpm in a 4-in. line and 175 gpm in a 6-in. line to achieve this velocity. At or below this flow, one risks the settling out or accumulation of solids in the line to a point where they may clog the pipe or possibly lock up or stall the pump. To avoid problems, manufacturers seek to restrict the use of a pump at low flow operation with enclosed impellers and, to a lesser extent, with vortex or recessed impellers. Line blockage, however, can occur regardless of impeller configuration.

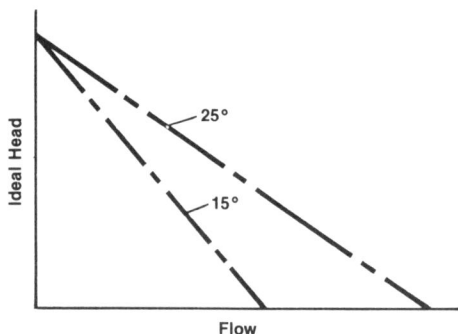

Figure 3-15: How vane exit angles affect pattern of flow.

To determine exactly where the pump will operate on the performance curve, the system head/capacity curve must be established and matched to the pump performance curve. **Figure 3-14** illustrates this. In System 1, the pump output flow would be X gpm at A head. In System 2, the pump output flow would be Y gpm at B head. System head curves are discussed in more detail later.

Differing pump hydraulic characteristics will result in one pump being better suited for a given application than another. The following paragraphs examine head-flow curves, pump performance curves, and variations in curve shape as a function of specific speed and of Pump Affinity Laws.

The head-flow curve of an ideal pump with ideal (frictionless) fluid is a straight line whose slope from zero flow to maximum flow varies with the impeller exit vane angle. For example, a given impeller and casing combination with an impeller vane exit angle of 25 degrees will have greater maximum flow than a similar impeller with a 15 degree vane exit angle. **(Figure 3-15)**

The actual head-flow characteristic of a centrifugal pump, however, is not an ideal straight line. Its shape is altered by friction, leakage, and the shock losses which occur in impeller and casing passages.

Friction losses in a centrifugal pump are generally proportional to the surface roughness and the wetted areas of the impeller

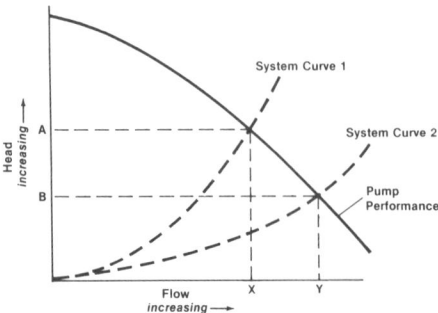

Figure 3-14: Matching system and pump curves.

Selection

and casing. *Leakage losses* result from the flow of liquid between the clearance of rotating and stationary parts, such as impeller to the case wear-ring. *Shock losses* occur as the liquid enters the impeller entrance vanes and as the liquid flows from the impeller into the casing.

These internal losses characteristically reduce pump performance from the ideal to the actual total head-flow curve shown in **Figure 3-16.** The flow at which the sum of all these losses is the least determines the point of maximum efficiency.

Mechanical losses in bearings, packings and mechanical seals further reduce pump efficiency. Although mechanical losses may be calculated, the results are generally not accurate; actual pump performance can only be determined by testing.

Design Considerations

Proper selection, operation and application of submersible centrifugal pumps require evaluation of a number of factors. These include net positive suction head (NPSH), cavitation, submergence, vortexing, and siphoning.

Net Positive Suction Head: NPSHA (net positive suction head available) is the absolute pressure of the liquid at the inlet of the pump. NPSHA is a function of elevation, temperature, and the pressure of the liquid, and is expressed in units of absolute pressure (psia).

The net positive suction head required (NPSHR) by a pump includes the velocity head at the suction flange, plus the head losses occurring in the inlet line between the flange and the impeller. Minimum NPSHR, where it is a critical operating concern, is specified by the manufacturer.

Submersible sewage pumps operate submerged in the liquid being pumped and are installed in wet wells vented to the atmosphere. The NPSHA is determined by the submergence measured from the inlet of the impeller to the surface of the liquid in the wet well, plus atmosphere pressure, minus the vapor pressure corresponding with the liquid temperature. In most cases, minimal submergence plus atmosphere pressure will result in an adequate NPSHA.

However, operating a pump with the following conditions present may result in an inadequate NPSHA — minimum submergence, flow rates higher than the Best Efficiency Point capacity, higher elevations, and/or elevated liquid temperatures. If these conditions exist, an analysis of the NPSHA in relation to the NPSHR is required.

Cavitation: Cavitation in a centrifugal pump can be a serious problem. Liquid pressure is reduced as the liquid flows from the inlet of the pump to the entrance to the impeller vanes. If this pressure drop reduces the absolute pressure on the liquid to a value equal to or less than its vapor pressure, the liquid will change to a gas and will form vapor bubbles. The vapor bubbles will collapse when the fluid enters the high-pressure zones of the impeller passages.

This collapse is called cavitation, resulting in a concentrated transfer of energy which creates strong local forces. These high-energy forces can destroy metal surfaces; very brittle materials are subject to the greatest damage. In addition to causing severe mechanical damage, cavitation also causes a loss of head, reduces pump efficiency, and can be noisy.

To prevent cavitation, centrifugal pumps must be provided with liquid under an absolute pressure which exceeds the combined vapor pressure and friction loss of the liquid in the area between the inlet of the pump and the entrance to the impeller.

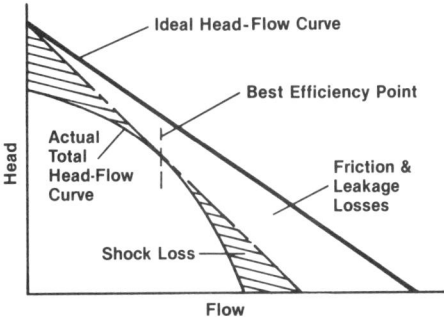

Figure 3-16: Factors in establishing Best Efficiency Point.

Selection

Submergence: A submersible centrifugal pump is cooled by the liquid surrounding it — the fluid it is pumping. To assure proper cooling, the manufacturer normally specifies recommended minimum submergence for continuous operation. In addition to the cooling function, the interaction of submergence with cavitation, vortexing, and siphoning can readily be seen.

Vortexing: This phenomenon results from pre-rotation of the liquid. Depending on depth of the pump setting, the vortex may become severe enough to reach to the surface, at which time air can be drawn into the pump, resulting in reduced performance and/or cavitation. If vortexing becomes a problem, it can be eliminated or minimized by increasing the minimum submergence level or by adding baffles at the pump inlet to break up the vortex.

Siphoning: If the ultimate level of the discharge line is below the level of fluid in the well, a condition termed "siphoning" can exist once the pump is started. Webster defines the term siphon as "A pipe or tube bent to form two legs of unequal length, by which a liquid can be transferred to a lower level, over an intermediate elevation, by atmospheric pressure forcing the liquid up the shorter branch of the pipe immersed in it, while the excess of weight of the liquid in the longer branch (when once filled) causes a flow."

The key words in the definition are "when once filled." The concern regarding a siphoning condition with a submersible centrifugal pump is that, once the pump is started, the siphoning will "assist" fluid flow and effectively reduce the head against which the pump must operate. If siphoning was overlooked in the selection or sizing of the pump, this assist can result in the pump overloading due to operation below its minimum recommended head.

Mechanical Design

A submersible pump is designed and constructed to operate submerged in the liquid being pumped. This eliminates the need for inlet or suction piping and also provides cooling for the motor. Obviously this makes the unit flood-proof.

The key components of a submersible pump are as follows.

Rotating Elements: Made up of the shaft, bearings, and impeller. The single shaft design of submersible pumps reduces the deflection and bearing wear associated with traditional flexibly-coupled designs. It also permits more rigid construction, thereby permitting operation at other than the critical speed of the shaft.

The bearings are either ball or ball/sleeve combinations, are permanently lubricated, and are positioned in the sealed motor cavity away from the fluid being pumped.

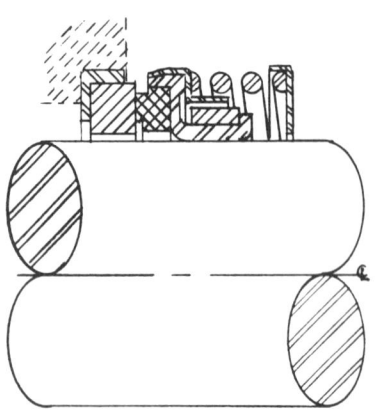

Figure 3-17

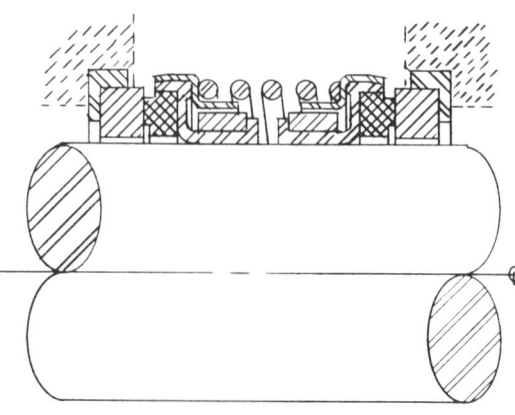

Figure 3-18

Selection

Mechanical Shaft Seals: The function of the seal in a submersible pump differs from that of a non-submersible, since it must protect against entry of liquid into the motor housing.

Submersible pumps are available with single, double, or tandem seals. **(Figures 3-17, 3-18 and 3-19)** Double or tandem-sealed pumps can be equipped with a seal failure sensor located in the oil reservoir between the seals. This arrangement offers early detection of lower seal failure and thus greater protection for the motor.

Pumpage, pressure, and speed each have their individual effect on the material, loading, and configuration selection of the shaft seals. These considerations interact with and upon each other to the extent that tolerances and finishes may need to be adjusted to meet a single specific application.

The pumpage has the single greatest effect on material selection. The standard combination of carbon-ceramic seals have successfully met a range of wastewater applications. For more severe applications and for longer service life, a wider selection of materials is now available. The pump manufacturer should be consulted for specific recommendations.

Pressure within the seal cavity must be considered to ensure that proper loading takes place at the seal faces. Too high a face load will cause wear, and too light a load may permit face separation and leakage. The pump and seal manufacturers must coordinate so that the proper sealing occurs over the entire normal pump operating range.

Speed is normally a minor consideration in seal selection for submersibles because the maximum speed is only 3600 rpm.

Motor and Motor Cavity: The submersible pump unit incorporates a liquid-tight motor cavity with proven field-serviceable motors, which are available in single or three phase designs and in a variety of common voltages. They range from approximately ½ to 100 hp or more, and are offered in oil-filled and air-filled versions.

Another important advantage of the submersible pump is the added heat dissipation obtained from the surrounding medium. In a non-submersible pump, the primary means of motor heat dissipation is by convection to the ambient atmosphere. Depending on installation, this can vary from frigid temperatures with maximum heat removal, to very high temperatures encountered on a summer day, made even hotter by a confined pit or building.

With a submersible pump, heat transfer is accomplished by direct conduction to the relatively constant temperature fluid being pumped. This more efficient heat transfer

Figure 3-17: Single seal. Left is the motor side and right is the water side. Figure 3-18: Double seal. Figure 3-19: Tandem seals. (Photos courtesy: Pac-Seal)

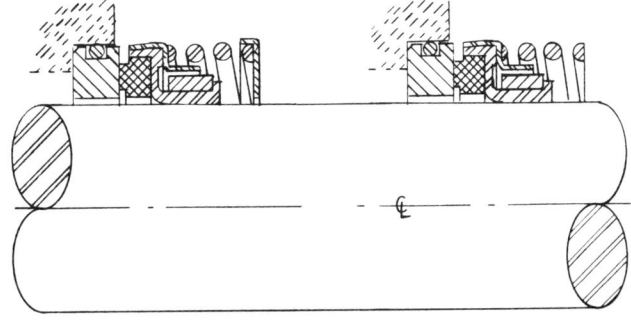

Figure 3-19

Selection

method results in a lower operating temperature for the motor and its internal components, and helps extend the life of the motor.

Wiring Cable Entry: The cable entry is designed to provide sealing around the O.D. of the cable. This permits complete pump submergence and provides strain relief to support the normal weight of the cable. The pump should never be lifted by the cable, since this strain relief is adequate for cord weights but will not support the weight of the pump.

Most cables used are rated SO, SJO, or STO, and are designed for extra-hard usage. These types of cable have been accepted by the NEC for use on submersible sewage pumps. Special attention is also given to jacketing compounds to prevent deterioration from the liquid being pumped or from ambient conditions.

System Losses And Head Curves

The total dynamic head (TDH) that the pump must overcome consists of two major elements — static head and friction head. *Static head* is the total vertical distance that the fluid must be lifted, from the lowest water level to the highest point in the discharge plumbing. *Friction head* results from the flow of the liquid through the discharge pipe, elbows, fittings, valves, etc. Each of these components is commonly expressed in feet (of water).

The total dynamic head of the system is determined as follows. Static head is the actual vertical distance measured from the minimum water level in the basin to the highest point in the discharge piping.

Friction head is the additional head created in the discharge system due to resistance to flow within its components. All straight pipe, fittings, valves, etc., have a friction factor which must be considered. These friction factors are converted to, and expressed as, equivalent feet of straight pipe, which can then be totalled and translated to friction head based on the flow and pipe size.

System Sizing

System sizing is based on three elements: (1) Data required, (2) factors to be determined, and (3) analysis of the pump operating point.

Data Required:

• Design flow (gpm) which the pump is required to produce at the total dynamic head (TDH) of the system.
• Static elevation of the system.
• Size of pipe being used.
• Type of pipe being used.
• Length of each size/type of pipe in system.
• C-factor of pipe (interior smoothness of pipe).
• Number and types of valves.
• Number and types of fittings.
• Sudden enlargements/restrictions in the system.

Factors To Be Determined:

• Target pump capacity. This figure can be derived from local plumbing codes, design or average flow tables, or fixture unit values. This is then equated to gallons per minute. All of these factors vary depending on the type of application and particular data.
• Friction losses within piping.
• Friction losses within valves and fittings.
• What type of pump will be utilized. The pump must be capable of developing the target flow in gallons per minute at a to-be-determined head above the static elevation of the system, plus any pressure to be encountered if pumping into a force main.

Establishing The Pump Operating Point:

• A table should be set up to determine the friction losses based on the parameters defined by the system.
• The flow should be chosen in increments which will provide a sufficient number of data points above and below the intersection of the system curve with the pump characteristic curve.

Selection

Table 3-20

EQUIVALENT LENGTH OF PIPE FOR VARIOUS FITTINGS*

FITTINGS	NOMINAL PIPE DIAMETER IN INCHES							
	1	1¼	1½	2	2½	3	4	6	8
STANDARD TEE STRAIGHT RUN	2	3	3	4	5	6	7	11	14
STANDARD TEE THROUGH SIDE OUTLET	3	4	5	6	7	8	11	16	20
SUDDEN ENLARGEMENT**									
d/D = ¼ ***	3	4	5	6	7	8	11	16	20
d/D = ½	2	3	3	4	4	5	7	10	13
d/D = ¾	1	1	1	2	2	2	3	4	5
GATE VALVE FULLY OPEN	1	1	1	2	2	2	3	4	5
45° ELBOW	2	2	2	3	3	4	5	8	10

*INFORMATION GATHERED FROM *THE HANDBOOK OF PVC PIPE, DESIGN AND CONSTRUCTION* BY THE UNI-BELL PLASTIC PIPE ASSOCIATION AND ROUNDED UP TO THE NEAREST FOOT.
**REFER TO d FOR THE COLUMN TO BE REFERENCED.
***THE NUMBERS ASSOCIATED WITH d/D = ¼ ARE ALSO VALID FOR A STANDARD 90° ELBOW OR A STRAIGHT RUN OF A TEE REDUCED ½.

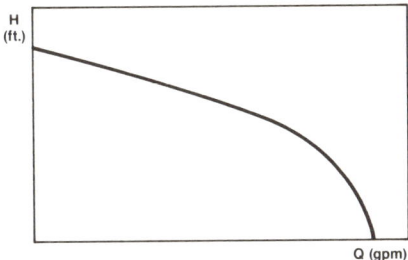

Figure 3-22: Pump characteristic curve — a function of the pump design.

• These losses can be determined for the pump size(s) and length(s) from the friction tables presented in the *Cameron Hydraulic Book, Hydraulics Institute Pipe Manual, Plastic Piping Systems Handbook*, etc., for the type of pipe being used.

• These values are usually given in friction loss or head loss per 100 ft. of pipe.

• The head loss per 100 ft. of pipe figure would then be multiplied by the overall length of that type of pipe in the system.

• Then the friction losses for the valves and fittings must be determined. These values are derived from tables presented in the manuals mentioned above.

• These values must be converted to friction loss in the equivalent length of straight pipe so it can be added to the overall pipe length in the system. (Refer to **Tables 3-20 and 3-21** for values.)

• Once all of these values have been determined, the system curve can be plotted on the same graph as the pump characteristic curve. This is illustrated in **Figures 3-22, 3-23 and 3-24**, and is typical of the procedure presented in most pump application books.

This procedure can also be used to analyze two pumps in equal parallel (a common discharge pipe with equal elevation and equal pipe lengths for each pump). The pump curve is adjusted to indicate double the capacity for a given head, and is then

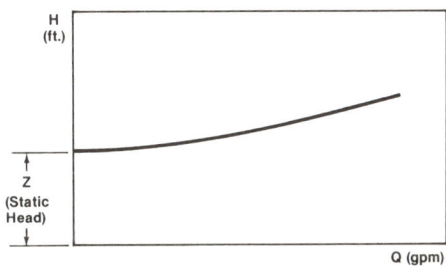

Figure 3-23: Piping system characteristic curve — a function of pipe, valve and fitting friction.

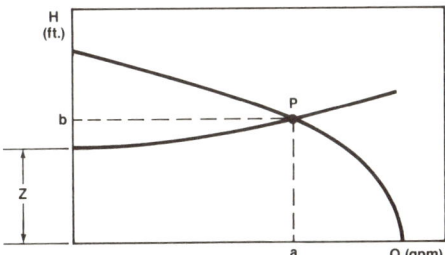

Figure 3-24: Matching pump performance and system loss curves.

Selection

TABLE 3-21
FRICTION LOSS IN SCHEDULE 40
PLASTIC PIPE

Velocity measured in ft./sec. Loss in feet of water head per 100 ft. of pipe.

GALS PER MIN	1/2" Vel	1/2" Loss	3/4" Vel	3/4" Loss	1" Vel	1" Loss	1 1/4" Vel	1 1/4" Loss	1 1/2" Vel	1 1/2" Loss	2" Vel	2" Loss	2 1/2" Vel	2 1/2" Loss	3" Vel	3" Loss	3 1/2" Vel	3 1/2" Loss	4" Vel	4" Loss
2	2.10	3.47	1.20	0.89																
4	4.23	12.7	2.41	3.29	1.49	1.08	.86	.27	.63	.12										
6	6.34	26.8	3.61	6.91	2.23	2.14	1.29	.57	.94	.26	.57	.09								
8	8.45	46.1	4.81	11.8	2.98	3.68	1.72	.95	1.26	.45	.77	.18	.52	.05						
10	10.6	69.1	6.02	17.9	3.72	5.30	2.14	1.44	1.57	.67	.96	.24	.65	.08	.43	.03				
12			7.22	24.9	4.46	7.72	2.57	2.02	1.89	.94	1.15	.37	.78	.11	.52	.05				
15			9.02	37.4	3.60	11.8	3.22	3.05	2.36	1.41	1.50	.52	.96	.17	.65	.07	.49	.05		
18			10.8	50.9	6.69	16.5	3.86	4.28	2.83	1.99	1.72	.70	1.18	.14	.78	.10	.58	.04		
20	5" PIPE		12.0	63.9	7.44	19.7	4.29	5.31	3.15	2.44	2.01	.86	1.31	.29	.87	.12	.65	.06	.52	.03
25					9.30	30.1	5.36	7.80	3.80	3.43	2.50	1.28	1.43	.43	1.09	.18	.81	.08	.64	.04
30	.49	.02			11.15	41.8	6.43	10.8	4.72	5.17	2.80	1.80	1.94	.61	1.30	.25	.97	.11	.77	.06
35	.57	.03			13.02	55.9	7.51	14.7	5.51	6.91	3.35	2.40	2.35	.81	1.52	.33	1.14	.15	.89	.08
40	.65	.04	6" PIPE		14.88	71.4	8.58	18.8	6.30	8.83	3.82	3.10	2.68	1.03	1.74	.43	1.30	.19	1.02	.10
45	.73	.04			16.70		9.65	23.5	7.08	10.9	4.30	3.85	3.02	1.32	1.95	.54	1.46	.24	1.15	.13
50	.82	.05	.57	.02			10.72	28.2	7.87	13.3	4.78	4.65	3.35	1.56	2.17	.65	1.62	.29	1.28	.16
55	.90	.60	.62	.02			11.78	33.8	8.66	16.0	5.26	5.55	3.69	1.88	2.39	.74	1.70	.34	1.41	.19
60	.96	.07	.68	.03			12.87	40.0	9.44	15.6	5.74	6.53	4.02	2.19	2.60	.90	1.95	.40	1.53	.22
65	1.06	.09	.74	.04			13.92	46.7	10.23	21.6	6.21	7.56	4.36	2.53	2.82	1.02	2.00	.47	1.66	.25
70	1.14	.10	.79	.04			15.01	53.1	11.02	24.9	6.69	8.64	4.69	2.91	3.04	1.21	2.27	.54	1.79	.30
75	1.22	.11	.85	.05			16.06	60.6	11.80	28.2	7.17	9.82	5.03	3.33	3.25	1.41	2.32	.60	1.91	.34
80	1.31	.13	.91	.05			17.16	68.2	12.69	32.0	7.65	11.1	5.36	3.71	3.49	1.54	2.60	.69	2.04	.38
85	1.39	.15	.96	.06			18.21	77.0	13.38	35.3	8.13	12.5	5.70	3.81	3.69"	1.66	2.72	.76	2.17	.42
90	1.47	.16	1.02	.87			19.30	84.6	14.71	39.5	8.61	13.8	6.03	4.61	3.91	1.92	2.82	.85	2.30	.47
95	1.55	.18	1.08	.07					14.95	43.7	9.08	15.3	6.37	5.07	4.12	2.04	2.93	.96	2.42	.53
100	1.63	.19	1.13	.08					15.74	47.9	9.56	16.8	6.70	5.64	4.34	2.33	3.25	1.05	2.55	.57
110	1.79	.23	1.28	.10					17.31	57.3	10.5	20.2	7.37	6.81	4.77	2.82	3.58	1.25	2.81	.61
120	1.96	.27	1.36	.18	8" PIPE				18.89	67.2	11.5	23.5	8.04	7.89	5.21	3.29	3.98	1.45	3.06	.80
130	2.12	.32	1.47	.13					20.46	78.0	12.4	27.3	8.71	8.79	5.64	3.81	4.22	1.68	3.31	.83
140	2.29	.36	1.59	.15	.90	.04			22.04	89.3	13.4	31.5	9.38	10.5	6.08	4.32	4.54	1.95	3.57	1.07
150	2.45	.41	1.70	.17	.96	.04			23.6		14.3	35.7	10.00	12.0	6.51	4.93	4.87	2.19	3.82	1.23
160	2.61	.46	1.80	.19	1.02	.05					15.3	40.4	10.7	13.6	6.94	5.54	5.19	2.47	4.08	1.37
170	2.77	.51	1.92	.71	1.08	.05					16.3	45.1	11.4	16.0	7.36	6.25	5.52	2.75	4.33	1.53
180	2.94	.57	2.04	.24	1.15	.06					17.2	50.3	12.1	16.8	7.81	6.58	5.85	3.07	4.60	1.70
190	3.10	.63	2.16	.26	1.21	.07	10" PIPE				18.2	55.5	12.7	18.6	8.24	7.28	6.17	3.39	4.84	1.88
200	3.27	.70	2.77	.28	1.28	.07					19.1	60.6	13.4	20.3	8.68	8.36	6.50	3.73	5.11	2.06
220	3.59	.83	2.44	.31	1.40	.08	.90	.03			21.0	72.4	14.7	24.9	9.55	10.0	7.14	4.45	5.62	2.44
240	3.92	.98	2.67	.41	1.53	.10	.98	.03			23.9	85.5	16.1	28.7	10.4	11.8	7.79	5.27	6.13	2.91
260	4.25	1.13	2.89	.47	1.66	.12	1.06	.04			24.9	89.2	17.4	33.0	11.3	13.7	8.44	6.07	6.64	3.28
280	4.50	1.30	3.11	.54	1.79	.13	1.15	.04					18.8	38.1	12.2	15.7	9.09	6.95	7.15	3.85
300	4.90	1.48	3.33	.62	1.91	.15	1.22	.05					20.1	43.2	13.0	17.9	9.74	7.90	7.66	4.37
320	5.13	1.66	3.56	.69	2.05	.17	1.31	.06					21.6	48.4	13.9	20.1	10.40	8.88	8.17	4.93
340	5.44	1.87	3.78	.76	2.18	.19	1.39	.07	12" PIPE				22.9	54.5	14.8	22.5	11.00	9.96	8.58	5.50
360	5.77	2.07	4.00	.86	2.30	.21	1.47	.07					24.2	60.2	15.6	24.9	11.70	11.0	9.10	6.15
380	6.09	2.28	4.22	.94	2.43	.23	1.55	.08	1.08	.03			25.6	66.7	16.5	27.7	12.3	12.2	9.59	6.58
400	6.44	2.5	4.43	1.03	2.60	.25	1.63	.09	1.14	.04			26.8	73.3	17.4	30.6	13.0	13.4	10.10	7.52
450	7.20	3.1	5.00	1.29	2.92	.32	1.84	.11	1.28	.05					19.5	36.7	13.9	16.7	11.49	9.31
500	8.02	3.8	5.56	1.36	3.19	.39	2.04	.13	1.42	.05					21.7	46.1	16.2	20.3	12.6	11.3
550	8.82	4.5	6.11	1.86	3.52	.46	2.24	.16	1.56	.06					23.9	55.0	17.9	24.3	13.0	13.5
600	9.62	5.3	6.65	2.19	3.85	.54	2.45	.18	1.70	.07					26.0	64.4	19.5	28.5	15.10	15.8
650	10.40	6.2	7.22	2.53	4.16	.63	2.65	.21	1.84	.09					28.2		21.1	33.0	16.40	18.3
700	11.2	7.1	7.78	2.92	4.46	.72	2.86	.24	1.99	.10							22.7	37.9	17.60	21.1
750	12.0	8.1	8.34	3.35	4.80	.82	3.06	.28	2.13	.11							24.4	43.0	18.90	24.0
800	12.8	9.1	8.90	3.74	5.10	.39	3.26	.31	2.27	.13							26.0	48.4	20.20	26.8
850	13.6	10.1	9.45	4.21	5.48	1.03	3.47	.35	2.41	.15							27.6	54.1	21.4	30.1
900	14.4	11.3	10.0	4.75	5.75	1.16	3.67	.39	2.56	.16										
950	15.2	12.5	10.5	5.26	6.06	1.35	3.88	.43	2.70	.18									22.7	33.4
1000	16.0	13.7	11.1	5.66	6.38	1.40	4.08	.48	2.84	.19										
1100	17.6	16.4	12.2	6.84	7.03	1.65	4.49	.56	3.13	.23										
1200	19.61	19.2	13.3	8.04	7.66	1.96	4.90	.66	3.41	.27										
1300	20.8		14.4	8.6	8.30	2.28	5.31	.76	3.69	.31										
1400	22.4		15.6	10.6	8.95	2.59	5.71	.88	3.98	.37										
1500	24.0		16.7	12.0	9.58	2.93	6.12	1.00	4.26	.42										
1600	25.6		17.8	12.6	10.21	3.29	6.53	1.12	4.55	.46										
1800			20.0		11.50	4.13	7.35	1.39	5.11	.57										
2000			22.2		12.78	5.03	8.16	1.69	5.68	.70										
2200			24.4		14.05	6.00	8.98	1.99	6.25	.85										
2400			26.7		15.32	6.7	9.80	2.37	6.81	.98										
2600							10.61	2.73	7.38	1.14										
2800							11.41	3.15	7.95	1.29										
3000							12.24	3.58	8.52	1.48										
3200							13.05	3.7	9.10	1.65										
3500							14.30	4.74	9.95	1.96										
3800							15.51	6.3	10.80	2.30										
4200									11.92	2.76										
4500									12.78	3.24										
5000									14.20	3.95										

*Date shown is calculated from Williams and Hazen formula $H = \frac{3.023}{C\ 1.852} \times \frac{V\ 1.852}{D\ 1.167}$ using C – 150. For water at 60° F.

Where H = head loss, V = fluid velocity ft./sec., D = diameter of pipe, ft.
C = coefficient representing roughness of pipe interior surface.

Selection

plotted with the system curve. Again, the intersection is the operating point of the duplex pump arrangement. (**Figure 3-25**)

This procedure can be extended for any number of pumps operating in equal parallel.

Figure 3-26 shows how individual pump capacity decreases and head increases as the number of operating pumps rise. This example of a system head and parallel pump curve illustrates centrifugal pumps of the same size operating in parallel. The result is a widely varying flow for few pumps operating simultaneously, or a narrowly varying flow for a larger number of pumps operating simultaneously.

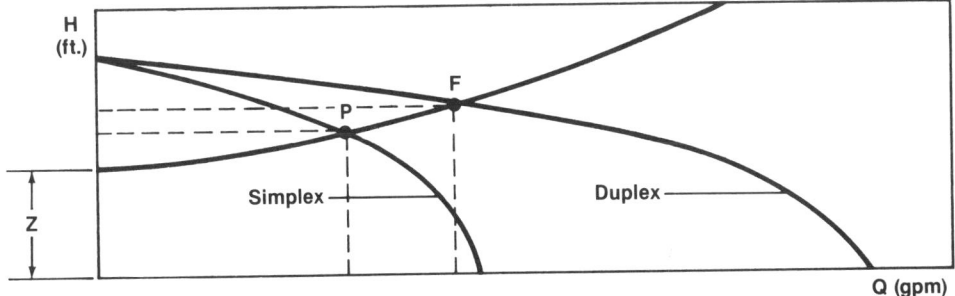

Figure 3-25: Illustration of change in system flow resulting from duplex operation.

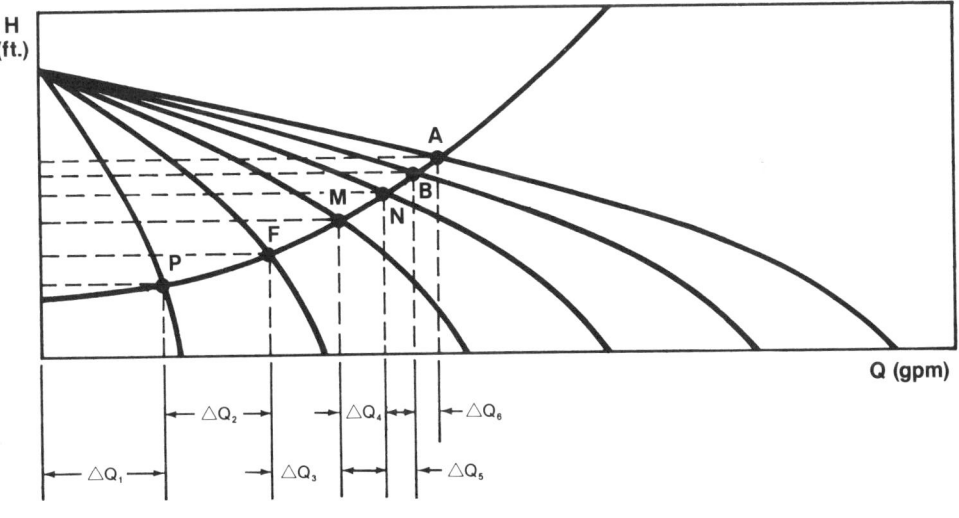

$\triangle Q_1 > \triangle Q_2 > \triangle Q_3 > \triangle Q_4 > \triangle Q_5 > \triangle Q_6$

Figure 3-26: System head and parallel multi-pump curves.

37

Selection

Submersible Pump Sizing

Most consulting engineers have a standard approach to lift station design. They consider and evaluate existing community needs, future development potential, local codes and regulations, elevations, force mains, etc., to arrive at a design flow.

Chapter 2 *(Sizing the Station)* discussed two methods of sizing a station handling flows of up to 3,000 gpm. One method discussed average daily flow, peak flow, etc., and proposed the formula:

$$Q \text{ in} = \frac{60\% \text{ of 24-hr. flow in gals.}}{4\text{-hr. period in mins.}} = \text{gpm}$$

and

$$Q \text{ in} = \frac{.60 \times h \times f}{4 \times 60} = \text{gpm}$$

This gives the gpm flow into the station, which could also be considered the minimum design pump gpm.

The second method discussed the number of homes, average number of people per home, and water usage per day, and presented a series of steps that could be converted to the formula:

(# of homes) (# of people/home) (average flow/person) = inflow/day

$$Q \text{ in} = \frac{(\text{inflow/day}) (\text{peak factor})}{\text{minutes/day}}$$

Either method will come up with about the same gpm.

Chapter 2 also discussed sump volume, with calculations to determine the minimum volume:

$$V \text{ min.} = \frac{(T \text{ min.}) (Q \text{ in})}{4}$$

At this point, we have determined:

A. The inflow into the station — or the design pump capacity in gpm.
B. The minimum effective sump volume in gallons.

The basic function of a submersible centrifugal non-clog pump is to pump a liquid with suspended solids *from* location and elevation A *to* location and elevation B. This assumes that A is at a *lower* elevation than B, and that A and B are connected by piping.

EXAMPLE PROBLEM

A consulting engineering firm has a contract to design a sewage lift station to move water from point A to point B. It elects to use submersible non-clog pumps. **(Figure 3-27.)**
A. Q in, the design pump capacity, has been calculated at 375 gpm.
B. The vertical *and* horizontal distance from point A to point B is 950 ft.
C. The vertical distance from A to B is 41 ft.
D. New cast iron will be used for the discharge pipe.
E. The engineer will design for a future condition, selected as 15 years from now.
F. The engineer will recommend a duplex station with each pump sized to handle the full design capacity.
G. To lift the water 41 ft. while moving it 950 ft., the water will pass through one tee, four 90-degree elbows, one check valve and one gate valve.

Selection

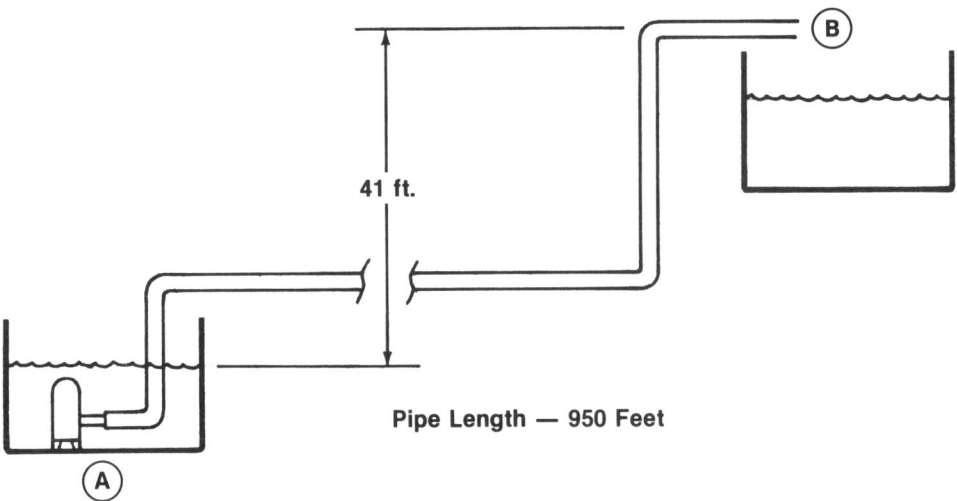

Figure 3-27: Example Problem situation.

Therefore:
A. Design pump capacity .... 375 gpm
B. Discharge pipe length .... 950 ft.
C. Static elevation .... 41 ft.
D. Pipe material used .... New cast iron
E. Design piping system for .... 15 yr.-old C.I. pipe
F. Type of system .... Duplex, two pumps
G. Fittings .... All cast iron (7 total)

When the pump turns On, it will push the water from point A to point B. It is necessary to size the piping system so that the pump will develop the capacity Q in (in gpm), figured previously at 375 gpm.

To calculate TDH:

1. Select a common pipe diameter — in this case, we will use 6-in. We must check for minimum velocity, which must be higher than two ft/sec, and for excessive velocity, which must be less than 10 ft/sec.

2. Calculate frictional loss, including design for old pipe and also for new pipe.

3. Add static head and friction head.

As noted earlier, the lower limit, 2 ft/sec, is considered the minimum flow to keep the solids moving so they don't settle out and clog the pipe. The upper limit, 10 ft/sec, usually results in an excessive frictional loss. In addition, since most sewage includes sand, grit or other abrasive material, this and higher velocities will sandblast the valves and fittings. (See **Table 3-28 — Charts A, B and C,** for 15-yr.-old pipe and fittings.)

To keep material costs down, we would first consider 4-in. cast iron pipe. At 375 gpm, both velocity and friction loss are high. At 400 gpm, Chart A shows a velocity of 10.21 ft/sec and a frictional loss of 16 ft. per 100 ft. This is excessive. Note that Chart C indicates that no adjustment to values is required, since the design is based on 15-year-old cast iron pipe.

With 6-in. pipe at 400 gpm, Chart A shows a velocity of 4.54 ft/sec and a frictional loss of 2.21 ft. per 100 ft. This is reasonable.

TABLE 3-28

FRICTION OF WATER IN PIPES AND FITTINGS
Williams & Hazen FORMULA C = 100
Loss of Head in Feet Due to Friction per 100 Feet of 15 Year Old Cast Iron Pipe

CHART A

Gallons Per Min.	4 Inch Pipe Vel.	4 Inch Pipe Fric.	6 Inch Pipe Vel.	6 Inch Pipe Fric.
40	1.02	0.22		
60	1.53	0.47		
75	1.92	0.73		
100	2.55	1.22	1.14	0.14
125	3.19	1.86	1.48	0.28
150	3.84	2.55	1.71	0.32
175	4.45	3.44	2.00	0.48
200	5.11	4.40	2.28	0.62
225	5.77	5.45	2.57	0.74
250	6.40	6.72	2.80	0.92
275	7.03	7.99	3.06	1.15
300	7.66	9.30	3.40	1.29
350	8.90	12.32	3.98	1.75
400	10.21	16.00	4.54	2.21
450	11.50	19.80	5.12	2.65
500	12.77	24.00	5.60	3.30
550			6.16	3.93
600			6.72	4.70
650			7.28	5.40
700			7.84	6.20
750			8.50	7.00
800			9.08	8.00
850			9.58	8.95
900			10.38	10.11
950			10.72	10.80
1000			11.32	12.04
1050			11.90	13.30
1100			12.50	14.31
1150			12.95	15.60
1200			13.52	16.69
1250			14.10	18.50
1300				
1400				
1500				
1600				
1800				
2000				
2200				
2400				
2600				
2800				
3000				

CHART C

Williams & Hazen Coefficient "C"	Multiplier to Adjust Chart	Pipe Description
100	1.00	Average 15 year old cast iron pipe
110	0.840	Vitrified sewer pipe
120	0.715	New wrought iron pipe
120	0.715	Average 5 year old cast iron pipe
130	0.615	Average new cast iron pipe
140	0.540	Very straight & smooth cast iron pipe
140	0.540	New steel pipe: Cement asbestos pipe
150	0.470	New CI pipe w/cent. spun bituminous lining
150	0.470	Schedule 40 PVC pipe

8 Inch Pipe Vel.	8 Inch Pipe Fric.	10 Inch Pipe Vel.	10 Inch Pipe Fric.	12 Inch Pipe Vel.	12 Inch Pipe Fric.	14 Inch Pipe Vel.	14 Inch Pipe Fric.	16 Inch Pipe Vel.	16 Inch Pipe Fric.
1.90	0.32								
2.20	0.42								
2.60	0.54	1.64	0.18						
2.92	0.68	1.80	0.21						
3.20	0.82	2.04	0.28	1.42	0.11				
3.52	0.97	2.25	0.33	1.57	0.14				
3.84	1.14	2.46	0.39	1.71	0.15				
4.16	1.34	2.66	0.46	1.85	0.19	1.37	0.09		
4.46	1.54	2.86	0.52	2.00	0.22	1.47	0.10		
4.80	1.74	3.06	0.59	2.13	0.24	1.58	0.11		
5.12	1.97	3.28	0.67	2.27	0.27	1.68	0.13		
5.48	2.28	3.48	0.75	2.40	0.31	1.79	0.14		
5.75	2.46	3.68	0.83	2.50	0.34	1.89	0.16		
6.06	2.87	3.88	0.91	2.70	0.38	2.00	0.18		
6.40	3.02	4.08	1.01	2.84	0.41	2.10	0.19	1.59	0.10
6.70	3.21	4.29	1.09	2.98	0.44	2.20	0.22	1.67	0.11
7.03	3.51	4.50	1.20	3.13	0.49	2.31	0.23	1.75	0.12
7.35	3.84	4.71	1.34	3.27	0.53	2.42	0.23	1.83	0.13
7.67	4.15	4.91	1.46	3.41	0.57	2.52	0.26	1.91	0.14
8.00	4.45	5.11	1.51	3.55	0.62	2.63	0.29	1.99	0.15
8.32	4.85	5.31	1.62	3.69	0.67	2.74	0.31	2.07	0.16
8.95	5.50	5.71	1.87	3.98	0.78	2.94	0.36	2.22	0.18
9.60	6.27	6.10	2.09	4.20	0.85	3.15	0.39	2.39	0.21
10.25	7.15	6.53	2.39	4.55	0.98	3.36	0.44	2.55	0.24
11.50	8.80	7.35	2.95	5.11	1.21	3.78	0.55	2.87	0.30
12.70	10.71	8.10	3.65	5.60	1.43	4.20	0.66	3.19	0.39
		8.98	4.24	6.25	1.81	4.60	0.81	3.51	0.46
		9.80	5.04	6.81	2.08	5.04	0.96	3.83	0.54
		10.61	5.81	7.38	2.43	5.46	1.13	4.15	0.62
		11.41	6.70	7.95	2.75	5.88	1.29	4.46	0.70
		12.24	7.62	8.52	3.15	6.30	1.47	4.79	0.80

Gallons Per Min.									
3200			9.10	3.51	6.68	1.67	5.12	0.88	
3400			9.66	3.91	7.10	1.86	5.44	0.98	
3600			10.25	4.37	7.52	2.08	5.77	1.10	
3800			10.80	4.90	7.95	2.36	6.07	1.20	
4000			11.35	5.32	8.40	2.47	6.38	1.34	
4500			12.78	4.70	9.45	3.22	7.20	1.65	
5000			14.20	6.40	10.50	3.92	7.96	2.02	
5500					11.55	4.65	8.78	2.39	
6000					12.60	5.50	9.56	2.60	

Velocities of less than 2' per second are not recommended for raw sewage.

CHART B

Length of Straight Pipe which will give same Friction Loss as listed Valves and Fittings

Pipe Size	Std. Elbow	Long Radius Elbow	45° Elbow	Tee through Side	Gate Valve Open	Swing Check Valve Open
4"	11'	7'	5'	22'	2.3'	27'
6"	16'	11'	7.7'	33'	3.5'	40'
8"	21'	14'	10'	43'	4.5'	53'
10"	26'	17'	13'	56'	5.7'	67'
12"	32'	20'	15'	66'	6.7'	80'
14"	36'	23'	17'	76'	8.0'	93'
16"	42'	27'	19'	87'	9.0'	107'

(Add to Actual Pipe Length to get Equivalent Length for use with Friction Factor from above Table).

Selection

With 8-in. pipe at 400 gpm, Chart A shows a velocity of 2.60 ft/sec and a frictional loss of only .54 ft. per 100 ft. This is also reasonable — but we should try the less expensive 6-in. pipe first.

Frictional Loss

Pipe length = 950 ft.
Fittings: Refer to Chart B — length of straight pipe which will give the same friction loss as listed. The result is **Table 3-29**, below.

Item	6" Pipe Friction Loss	Quantity	Total
Tee	33	1	33
90-degree Elbow	16	4	64
Check Valve	40	1	40
Gate Valve	3.5	1	3.5
Total:			140.5 ft.

Total equivalent length — what the pump "sees:"

Pipe length   950
Fittings   140.5
               1090.5 ft.

Friction Head: Refer to Chart C. From Williams & Hazen, C = 100. This is our *design for con-dition* — what is going to be the loss based on friction after 15 years' use. The chart is based on 15-year-old cast iron pump, and no further adjustments or corrections are needed.

Proceed to Chart A. Since 375 gpm is not shown, select the next highest value shown. In this example, this value is 400 gpm. The friction loss at 400 gpm in 6-in. pipe is 2.21 ft. per 100 ft. The velocity is 4.54 ft/sec, which is satisfactory.

(2.21'/100") (1090.5') = 24.10' (friction loss)

Therefore, static head   = 41
          friction head = 24.1
          TDH   = 65.1 ft.

The approximate design future conditions are 400 gpm at 65.1 ft. TDH.

We can estimate the brake horsepower by using the formula and assuming a pump efficiency (EP). A reasonable efficiency to assume is 55 percent.

$$BHP = \frac{(gpm)(TDH)}{(3960)(EP)}$$

$$BHP = \frac{(400)(65.1)}{(3960)(.55)} = 11.95$$

Therefore, we are looking for a pump in the 10 to 15 hp range.

To determine the actual head at 375 gpm, we have to plot a *system curve,* or system head curve.

To do this, we first must tabulate:

1. Various gpm flows — from zero to something higher than 375 gpm.
2. The velocity at each flow.
3. The frictional loss or head at each flow.
4. Add the frictional and static head — which equals TDH.

Selection

5. Plot the curve.

In this example, we will vary gpm from shutoff (zero) to 550 gpm.

We will use 6-in. cast iron pipe. See Chart A, which tabulates velocity and friction for 15-year-old cast iron pipe. This is what Williams and Hazen set up as a standard, calling it coefficient C = 100.

Using this data, we prepare **Table 3-30** below. We can now plot the curve, with TDH on the vertical axis and gpm on the horizontal axis.

GPM	Velocity Ft./Sec.	Friction/100' C = 100	Friction per 1090.5 Feet	Static Head	TDH
0	0.00	.00	0.00	41	41.00
100	1.14	.14	1.53		42.53
150	1.71	.32	3.49		44.49
200	2.28	.62	6.76		47.76
250	2.80	.92	10.03		51.03
300	3.40	1.29	14.06		55.06
350	3.98	1.75	19.08		60.08
400	4.54	2.21	24.10		65.10
450	5.12	2.65	28.89		69.89
500	5.60	3.30	35.98		76.98
550	6.16	3.93	42.86		83.86

This is the system head curve, **Figure 3-31**. It shows the head required to make a liquid flow through a system of piping, valves, elbows, etc. If we change any variable in the system, the slope of the curve will also change.

From the curve, 375 gpm will require 63 ft. TDH.

Most system engineers have access to pump manufacturer's catalogs, so at this point the engineer selects or has determined (1) design solid size and (2) maximum allowable rpm.

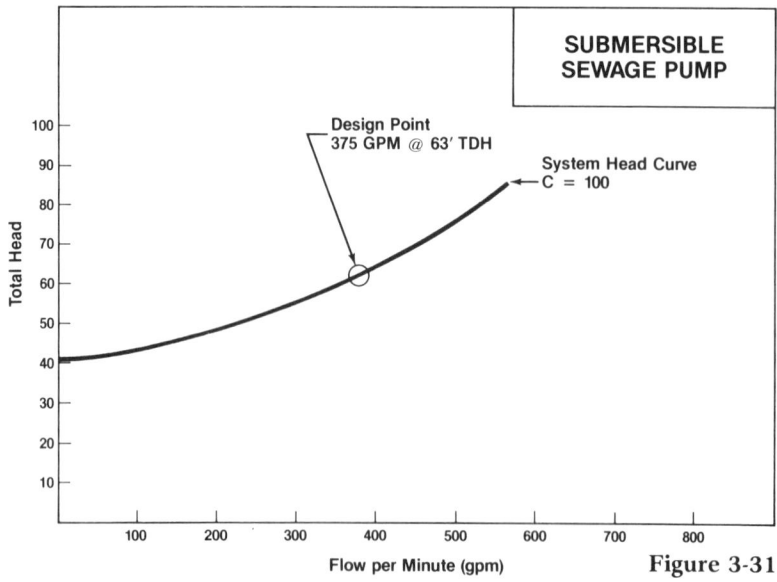

Figure 3-31

Selection

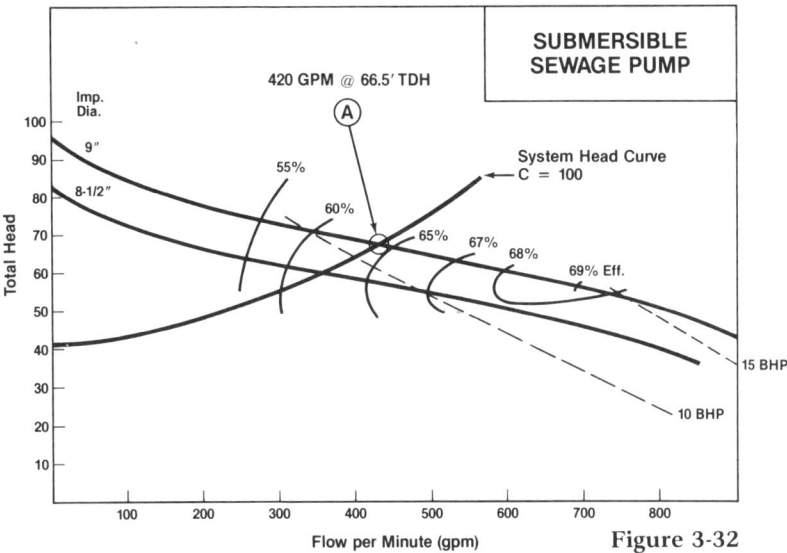

Figure 3-32

The next step is to plot the system head curve on a typical manufacturer's performance curve for a 10 to 15 hp pump. (**Figure 3-32**) This may be done simply to be certain that one or several manufacturers can quote the pump. Or the engineer may select a specific model and request a quotation on it "or equal."

Shown on the figure is pump performance with impeller diameters of 8½-in. and 9-in. Also shown are lines of constant efficiency and horsepower.

The pump selection could be a 15-hp unit with a 9-in. impeller. The performance is 420 gpm at 66.5 TDH. This is point A.

The engineer should now calculate a new system head curve based on what flow and pressure will *occur at start-up* — with new cast iron pipe, C = 130.

By the same method as before, but this time using the coefficient C = 130 multiplier of .615 on column 4 (friction per 1090.5 ft.), we can calculate the frictional loss on average new cast iron pipe (**Table 3-33,** below).

GPM	Friction per 1090.5 Feet C = 100		C = 130 Mult.		New Friction Per 1090.5 Feet		Static		TDH
0	0.00	times	.615	=	0.00	+	41	=	41.00
100	1.53				.94				41.94
150	3.49				2.15				43.15
200	6.76				4.16				45.16
250	10.03				6.17				47.17
300	14.06				8.65				49.65
350	19.08				11.73				52.73
400	24.10				14.82				55.82
450	28.89				17.77				58.77
500	35.98				22.13				63.13
550	42.86				26.36				67.36

We can now establish a new system head curve — **Figure 3-34.**

Selection

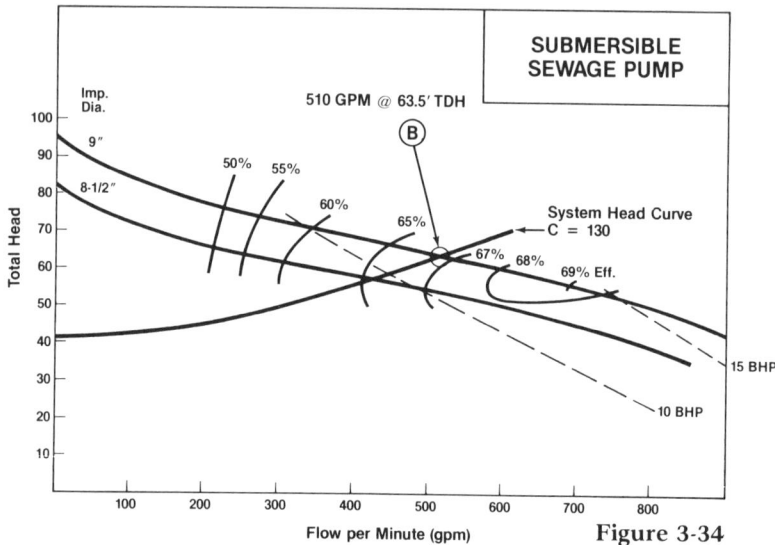

Figure 3-34

This system head curve crosses the 9-in. impeller performance curve at point B, or 510 gpm at 63.5 TDH. Pump efficiency is now 66.5 percent and calculated BHP is 12.3.

The selection would still be a 15 hp pump.

However, if the engineer specifically requests that the pump deliver the originally specified 375 gpm at 63-ft. TDH, the manufacturer could trim the 9-in. pump impeller to achieve this condition. This is point C on **Figure 3-35.**

With this new impeller trim of approximately 8-5/8 in., the new pump efficiency is 63 percent and the new BHP is 9.47. This is the future design condition.

The pump size could now be reduced to 10 hp.

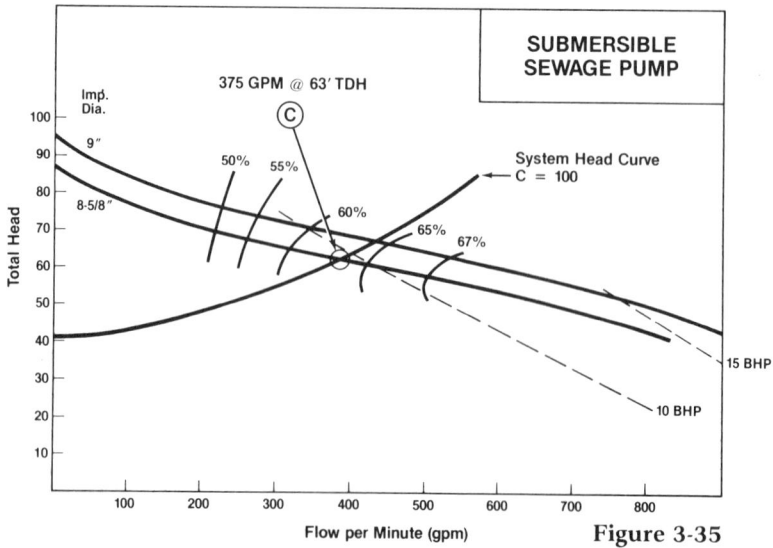

Figure 3-35

Selection

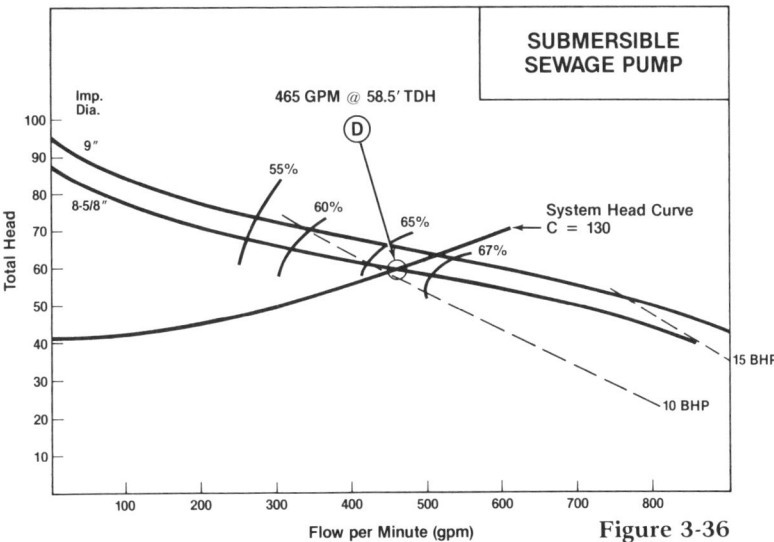

Figure 3-36

If a 10 hp pump is selected, installed with new cast iron pipe (C = 130), and equipped with the 8-5/8 in. impeller, it will operate at start-up at point D **(Figure 3-36)** and deliver:

465 gpm at 58.5 ft. TDH.
New pump efficiency is 66 percent.
New BHP is 10.4.

This exceeds the pump nameplate rating. The pump will also draw more amperage than shown on the nameplate. In this example, the horsepower overload is minimal — although it could be significant in other situations. Both present and future conditions should be checked, and overloading situations avoided.

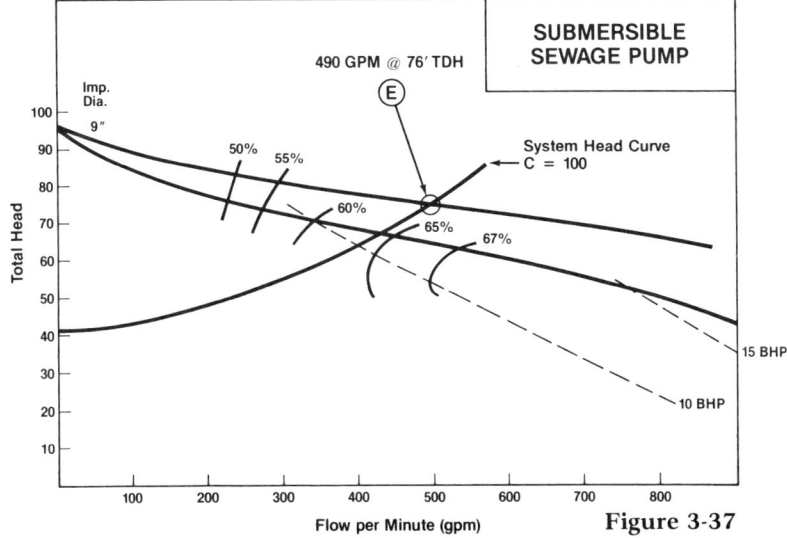

Figure 3-37

45

Selection

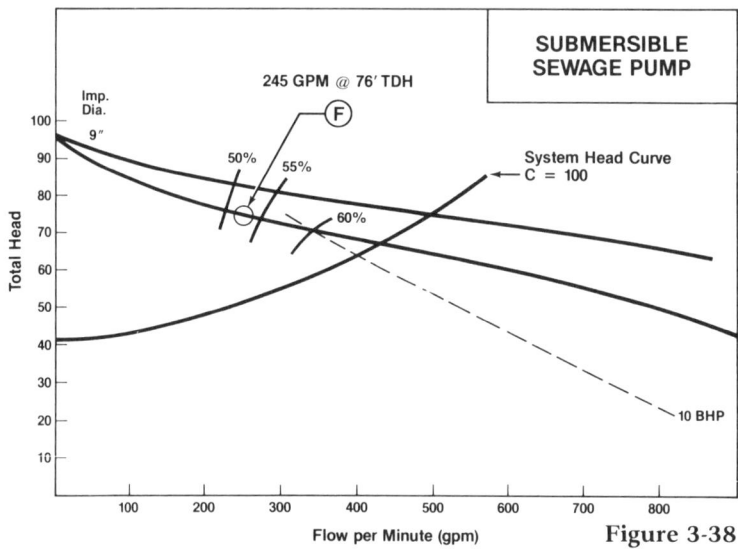

Figure 3-38

Duplex Sizing

Figure 3-37 illustrates duplex pump performance with one system head curve. We have selected the system head curve, C = 100, as plotted previously, and are using two 15 hp pumps with 9-in. impellers operating in parallel.

Duplex pump performance is plotted by doubling the single pump's capacity at the same head. This condition holds true only if both pumps are discharging through a common main.

Point E shows duplex pump performance — 490 gpm at 76 ft. TDH.

Point F on **Figure 3-38** shows that each pump will deliver half the capacity, 245 gpm — but at 76 ft. TDH.

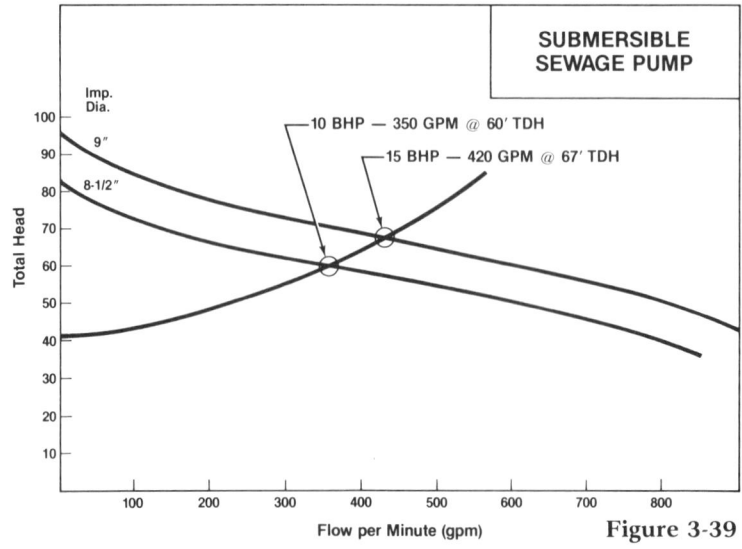

Figure 3-39

Selection

Please remember that, with both pumps operating, the capacity of each pump is reduced — and therefore pump efficiency and brake horsepower are also reduced.

In this example, pump efficiency is 51 percent and brake horsepower is 9.2.

It is sometimes desirable to install two or more different sized pumps in the same well.

Figure 3-39 illustrates pump performance with a 10 hp pump with an 8½-in. impeller and a 15 hp pump with a 9-in. impeller.

If they are operated singularly with the system head curve as shown, the 10 hp pump will deliver 350 gpm at 60-ft. TDH and the 15 hp pump will deliver 420 gpm at 67-ft. TDH.

If they are operated in parallel — discharging through a common main — the combination pump performance curve is drawn by adding the smaller pump's capacity to the larger pump's capacity, but only at the same head.

1. Transfer the zero flow, shut-off head, of the smaller pump — point A — to the larger pump's performance curve at the same head. Note **Figure 3-40.**
2. At the same head — assume 75-ft. TDH — add the capacity (65 gpm) of the smaller pump to the capacity (250 gpm) of the larger pump.
3. Continue adding capacities, at equal heads, until there is a locus of points.
4. Connecting these points results in a duplex performance curve of two unequal sized pumps running in parallel.

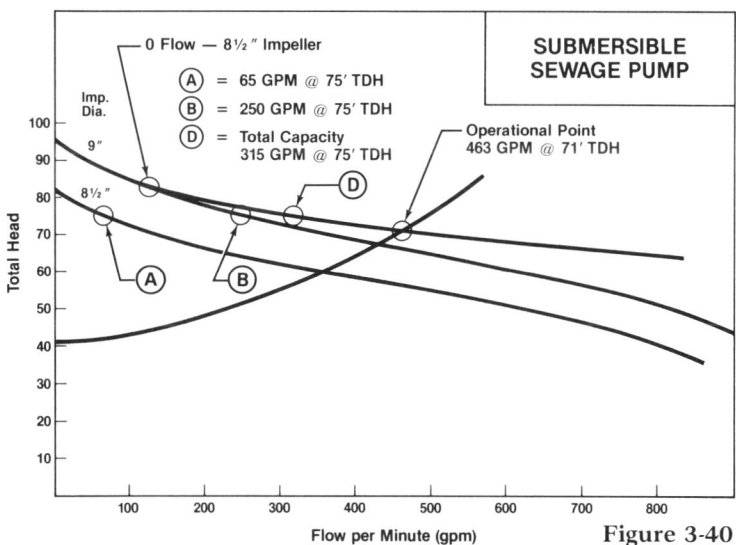

Figure 3-40

Other Considerations

The consulting engineer should also consider the following.

Pump flexibility — Is this the largest pump the manufacturer makes in this model? What could the user do if he has to pull this pump out in a few years? Can it be replaced with the same pump model but with a larger impeller and higher horsepower?

Selection

Pump history — Is this a new or a tried and tested unit? Can the user depend on getting parts 15 to 20 years from now?

Pump service — Does the manufacturer have many of these units in operation? Do they have a good record? Is local service available? How much factory help can we expect?

Pump performance — Has the pump selected for this application been sized to operate at or near the best efficiency point? As time goes by, will the pump decrease or increase in efficiency? Is there a possibility of operation at minimum head conditions? If so, will the pump draw excessive current and overload the motor? Can we obtain motor efficiency and/or input KW to calculate wire-to-water or overall efficiency — and thus calculate operating cost?

Summary and Reminders

System curves can be expanded to incorporate one curve at minimum sump level and the second curve at maximum sump level. This could become very important if the distance between the minimum and maximum levels becomes great.

Where two or more pumps will discharge into a common header pipeline, it is usually advantageous to ignore the head losses in the individual pump discharge (called station losses) from the head capacity curves. This is advisable because the pumping capacity of each unit will vary depending on which units are operating in parallel. (A further discussion can be found on pages 57-59 of the *Design of Wastewater and Stormwater Pumping Stations, Manual of Practice: 1981.*)

In systems where there is high velocity, *velocity head* must also be considered. It is an additional pressure created by the high velocity of the fluids moving in the piping system which is not indicated on the pressure gauge. This head must be figured into the total dynamic head of the sytem.

Velocity head can be determined by:

$$h_v = \frac{.00259(Q)^2}{d^4}$$

Where: h_v = velocity head
Q = gpm
d = inside diameter of circular pipe (inches)

This formula is found on pages 3-6 of the *Cameron Hydraulic Book.* Friction tables detailing velocity head applications can be found in *Cameron* on pages 3-12 and 3-47.

A pump station is normally designed to accommodate existing conditions with a capacity adequate for a specific time period in the future. If the anticipated growth is large, an increase in capacity may be required.

Options here include changing either the pump impellers or replacing the pumps with larger capacity units. The station could utilize an oversized discharge pipe if care is taken to maintain the velocity at or above 2 ft/sec.

Another option is to lay parallel lines during initial construction. This permits both force mains to be used when the future capacity is reached.

Final Example Results

After the above analysis, you have:
• determined the actual operating conditions of the pump system in gpm and TDH
• selected a pump which operates at the best possible efficiency point and utilizes the minimum horsepower required to operate in a non-overloaded condition
• determined whether the scouring velocity is within acceptable limits based on the type of wastewater being handled — normally 2 to 5 ft/sec.

You have also determined whether (if any parameter is not within acceptable limits) you need a larger capacity pump, or a pump with higher pressure capabilities, or smaller or larger pipe, or a different type of pipe, or a different piping arrangement.

This system curve analysis gives the design engineer and the manufacturer of the equipment confidence about how the pumps will operate within the system, and that the proper equipment has been selected for the application.

Chapter 4
Submersible Pump Controls

The control system is the "brain" of the submersible wastewater pump station. Since most stations are in unattended locations, remote from central sewage plants, each station must function automatically and reliably. Control systems should be designed to perform as simply as possible, while ensuring that the specific pumping job is done properly. They should incorporate such basic features as a disconnect for each pump, short circuit and overload protection in each phase of operation, and pump run indicators. The control panel should be easy to maintain and troubleshoot. This chapter discusses power supply, motor controllers, types of level controls, alarms, auxiliary system components, and control panel enclosures.

Power Supply

Primary Power Sources

Determination of the available primary or normal power supply at the site selected for a pump station is the first requirement. The local power company usually establishes the type of power source available and, depending upon the anticipated revenue from the installation, may consider accommodations not presently available.

Power sources are of two general types:
Single Phase — 2 or 3 wire
Three Phase — 3 or 4 wire

Two-wire single phase systems are generally 208 or 240 volts, have no neutral, and require a transformer for 115 volt power. In rare instances, a two-wire system may be 115 volt, where one wire will be a neutral. Three-wire 208 or 240 volt systems have two power wires and a neutral. One leg to neutral provides 115-120 volt power.

Three phase power supplies are 208, 240, 480 or 575 volt. Three-wire systems have no neutral and will require transformers for 115-120 volt power. Four-wire systems have neutrals and, in the case of 208 and 240 volt, are inherently capable of supplying 115-120 volt between at least one phase and neutral. If there is no independent 115-120 volt supply, transformers will be required for 480 and 575 volt systems.

Three phase systems are classified as *Wye* or *Delta*, referring to the method of connecting the supply transformers. In general, 240 volt systems are Delta, while 208, 480 and 575 volt systems are commonly Wye.

On all Delta systems, the voltage between one of the three phases and the neutral is much higher (180 to 210 volts) than the normal 115-120 volts; in these cases, unless a transformer is used, only two phases are available for 115-120 volt power requirements. Since this type of power supply is susceptible to voltage unbalance, phase monitoring circuitry is recommended.

A service entrance-rated main disconnect is recommended as good practice, although it is not always required by code. It may be a fused or unfused disconnect switch, a main circuit breaker in a separate enclosure, or built into the control panel. These devices are available in a variety of NEMA-rated enclosures to meet nearly any installation requirement.

Standby Power Sources

A standby or secondary power source may be required by the approving authority. It generally is an engine-driven generator

Electrical

set providing the same characteristics as the primary power source.

If the generator set (diesel, gasoline or propane) is permanently installed at the lift station, it usually includes a transfer switch which automatically starts the generator and transfers to standby power in the event of a primary power failure.

If a portable generator set is used, two additional components are required. These are a standby power receptacle to provide a quick, safe way of connecting the standby source, plus a manual transfer switch to disconnect the normal power source and reconnect the load to the standby power. The manual transfer switch also functions as an isolator to prevent cross-connection of the two power sources. Care must be taken to assure that the receptacle and mating plug are compatible.

It is sometimes desirable to incorporate these components into the main pump control panel, which can be done most simply with two standard circuit breakers with interlocks or a horsepower-rated rotary switch; this arrangement is more desirable to some users than commercially available but bulky double-throw switches. It also serves as a main breaker for the standby power source in the event the portable generator is substantially larger than required.

Whether a permanently installed or portable generator is used, the size of the power source required can be reduced by including a lock-out circuit in the control panel so only a selected number of pumps can operate in the emergency mode.

Motor Controllers

Across Line Starters

The most common form of motor controller is the across the line, full voltage, non-reversing (FVNR) magnetic starter. NEMA standards for these devices have been established to provide uniformity of size and horsepower ratings, and all are equipped with overload relays. They are available in several NEMA-rated enclosures (see listing at end of this chapter), as well as for open installations. If an across line starter is to be used as a stand alone, in conjunction with a pilot type control panel, a combination starter is generally specified in an enclosure type suitable for the environment. It includes the required short circuit protection (circuit breaker or fused disconnect) as well as overload protection.

Contactors may be used in most applications in place of NEMA-rated starters; they will function equally well when sized to accommodate the motor in-rush current. Since most contactors are either horsepower or ampere-rated, they should not be used for induction motors with full load amperage above NEC tabulations.

To comply with the code requirements for overload protection, a separate overload relay is required when contactors are used to start polyphase motors. Since many motor manufacturers include overload protection as an integral part of single phase motors, separate overload control devices may not be required. Contactors in conjunction with overload relays provide an economical and satisfactory answer to motor control in most lift station applications.

Overload provisions for induction motors are required by the NEC. When not included as part of the motor, as in polyphase motors, the overload protection generally takes the form of an overload relay equipped with one or more heater elements. Current flows to the motor through these elements. If this current rises above the trip setting, the relay contacts are opened — interrupting current to the starter or contactor coil, thus de-energizing this device and stopping current flow to the motor.

To conform to NEC, overload relays must have heater elements in all three power legs of three phase motors. Overload relays for single phase motors, if required, need only one heater element unless a particular manufacturer requires two elements for warranty purposes.

Heater elements are generally of two types, melting alloy or bimetal. The ambient compensated, quick-trip type bimetal heater element is recommended for sub-

Electrical

Photo 4-1: Exterior view of standard electrical control panel.

mersible pump motor applications; the melting alloy type may react too slowly to prevent motor damage. In addition to protecting against physical motor overload, this type of relay provides protection against low voltage.

Short circuit protection is required by NEC. This usually takes the form of a fused disconnect, standard circuit breaker, or an *MCP* (motor circuit protector). Of the three types, a standard circuit breaker is the most popular because it can be re-set, does not require a stock of proper fuses, normally meets code requirements for a disconnect means at the motor site, and is usually the most economical.

A standard circuit breaker in conjunction with NEMA-rated starters or the contactor/overload relay combination generally provides the simplest means of integrating the motor controllers with the level control system, using a single control panel in an enclosure suitable for the environment.

The MCP is a special form of circuit breaker developed specifically for induction motors. It has no thermal overload capability. It is often used for higher horsepowers (25 hp and up) and higher voltages where its cost is competitive.

Ground Fault Interruptors

Circuit breakers with *(GFI)* ground fault interruption are available from most manufacturers and are intended primarily for personnel protection. They will trip when a current flow of approximately 5 milliamperes to ground occurs on the load side of the circuit breaker.

They are used primarily with single phase convenience outlets, up to 30 ampere capacity. They offer protection to personnel when they are using portable drills, trouble lights, etc. Electrical codes usually require GFI protection for these applications.

While GFI protection for pumps is available, it is not required by NEC for sewage wet wells. If used for this type of service, nuisance tripping of the pumps may occur because of the normally wet environment. When GFI protection is used, special consideration should be given to locating all electrical connections in a dry environment.

Ground fault protection for pumps uses current sensors coupled with a solid state relay which has normally open or normally closed contacts. The relay output may be used to trip the pump circuit breaker, provided a shunt trip device is included, or it may be incorporated in the pump pilot control circuit to de-energize the pump starter.

The current sensors are available with selectable trip ranges to provide any desired equipment protection level. This enables the selection of ranges sufficiently high to prevent nuisance tripping while providing adequate equipment protection. This type of GFI pump protection can substantially increase the cost of the system.

Reduced Voltage Starters

Reduced voltage starters are a means of limiting the motor (starting) in-rush current. They are required when it has been determined — usually by the utility — that the power supply at a given site is limited in the amount of induction motor in-rush current the primary system can handle. Three types are in general use for centrifugal pump motors: auto-transformer type, Wye-Delta starters, and step starters.

Electrical

Auto-transformer type starters are the most popular since they can be used with all standard motors and without special windings. The starter consists of a three-phase auto-transformer (usually with three standard reduced voltage taps — 50%, 65%, 80%), three contactors, a time delay relay and an overload relay. This is called a *closed transition* starter.

Care must be exercised in the use of this type of starter since the in-rush current of the transformer may meet or exceed the motor in-rush across the line, depending upon system voltage, motor horsepower and motor starting code (see NEC). It is usually the most expensive of the reduced voltage types and requires the most installation space.

Wye-Delta starters take their name from the manner in which the motor windings are connected — the windings are connected in Wye to start the motor and then reconnected in Delta to run.

Variable Speed Controllers

Variable speed controllers are normally used to maintain pump flow rates approximately equal to rate of flow entering the lift station. While their application to submersible pumps is limited, the major types are discussed briefly below.

Multi-speed starters are available in several types, but the two-speed starter for a one winding constant or variable torque motor is the most common in pump applications. The two-speed starter consists of two magnetic starters, each with overload relays and auxiliary contacts to prevent both starters from being energized simultaneously.

The level control system must be designed to start two-speed motors in the desired sequence — with the limitation that a motor running at high speed must be stopped before restarting at low speed. For automatic operation, this is usually accomplished by the addition of compelling relays in the control circuit. Because six motor leads are required, it is not commonly used for submersible applications.

Variable frequency drives can be applied to conventional polyphase motors. They achieve speed regulation by controlling frequency, usually between 50 and 60 Hertz. A transducer, sensing discrete changes in wet well level, sends a signal to the variable frequency controller to raise or lower the frequency, thus speeding up or slowing down the motor.

Control circuitry is solid state and electronic starters are generally used. Conventional starters or contactors may be required with some units depending upon UL, NEC and local code requirements and on the specific application.

The variable frequency controller consists of a rectifier, a "chopper" and an inverter. The AC input is first converted to DC by the rectifier and then put through the chopper, at which point the control logic input adjusts the frequency and inverts it back to AC for motor input. Manufacturers claim the highest operational efficiency of any variable speed controller for the variable frequency units. It can accept a variety of input logic devices. These units are relatively high in cost, and require skilled maintenance personnel and sophisticated test equipment for trouble-shooting.

Level Control Systems

There are six devices available for wet well level control: (1) mercury float, (2) diaphragm type, (3) bubbler (air) systems, (4) electrode level sensing, (5) ultrasonic level detectors, and (6) pressure transducer. Each of these devices has its own virtues in submersible pump applications. Selecting one over another depends on the particular application.

Mercury Float Switches

The most popular level control system for submersible pump stations is the encapsulated mercury switch. It is suspended in the wet well at the desired actuation levels by a control cable which connects the switch to the control panel.

Simplex pump systems often use three switches — one for On, one for Off and a third for alarm, when used. Or they may use a single differential float switch for On-Off control.

Electrical

Multi-pump systems with a common Off level require one more switch than the number of pumps — i.e., a three pump system requires four float switches. Many other combinations can be used if the application involves separate Off levels, high and low speed pump operation, four or more pumps, duplex wet wells, etc.

Three float switches are required for a standard duplex system — the lowest for pumps Off, the next highest for lead pump On, and the third at a still higher level for lag pump On and the alarm function.

The float switches are normally open. As the wet well level rises, the Off float switch closes but no other action occurs. As level continues to rise, the lead On switch closes, completing the pilot circuit to start the lead pump. When the lead pump lowers the level and the lead On switch opens, no action occurs due to a latching contact which keeps the circuit energized until the level is lowered to the point at which the pump Off switch opens.

Float switches are maintained at the desired level by a weight attached to the control cable or by a swivel fitting which secures the switch to a pipe or other support. The mercury switch is usually encapsulated in polyurethane foam or enclosed in a solvent welded plastic shell. This provides flotation, seals the unit against the entry of water, and maintains electrical integrity.

The differential float switch, where one switch provides both On and Off actuation at different levels, is often used in applications that require a 6 to 30-in. range of function. Most such switches available have a minimum differential level of approximately 6-in. and a maximum differential of 30-in.

Mercury float systems are relatively simple to maintain. The mercury switches can be replaced easily, and highly skilled personnel are not usually required for troubleshooting.

Diaphragm Type

Diaphragm type switches are a special form of pressure switch. They consist of a

Photo 4-2: Interior view of electrical control panel.

housing containing a set of spring-loaded contacts, designed for snap action, which are actuated by a flexible diaphragm, usually neoprene. The switch must be mounted in a fixed position in the wet well approximately 10 to 12 inches below the desired actuation level, since this amount of water pressure is required to achieve the snap action in most designs. The electrical cord on these switches also contains a small diameter plastic tube which is required for venting of the switch interior to atmosphere so that predictable actuation levels can be maintained.

This type of switch provides a fixed differential — usually 10 to 12 inches — between switch open and switch closed positions. This differential often can be utilized to advantage in specific applications. While some designs provide greater or smaller differentials, none of the switches are adjustable.

Diaphragm type switches function in a control system exactly the same as float switches. However, because of the possibility of punctured diaphragms or plugged or pinched breather tubes, they are less commonly used than float switches.

Bubbler Systems

Bubbler or air systems are level controls which sense water level by creating a back pressure in a pneumatic system which is proportional to the depth of water submerging the end of the bubbler tube. An air supply is usually provided by an air compressor

Electrical

with a small receiver, a pressure regulator to control downstream pressure, a flowrator (needle valve in graduated flow tube), and a differential pressure regulator (across flow tube, inlet to outlet).

Pressure switches connected to the bubbler tube, on the discharge side of flow tube, are used to set desired actuation points for pump start, stop, and alarms. Pressure gauges are graduated in feet or inches of water as desired. They are connected to the bubbler line and provide a continuous visual indication of water level. Switch gauges are available which combine both gauge and switch features; there are two set points per unit which can be readily adjusted from the front of the instrument.

Since a controlled air supply is critical for proper operation of a bubbler system, it may be desirable to use duplex alternating air compressors with a low air supply alarm to ensure system reliability. Some systems available use vibrating reed or diaphragm-type compressors; these have no receiver and run continuously. Although less costly than piston-type units with receivers running cyclically, this type is less reliable. The air receiver also has the advantage of providing a volume of stored air which can both maintain air flow during brief power outages and provide necessary volume for purging.

Another feature considered necessary in a quality bubbler system is the differential pressure regulator installed across the flow tube (also called a minirator). A bubbler system depends upon a constant, low rate of flow (cubic inches per hour) to accurately reflect water level through back pressure in the air system. As water level increases and decreases, the differential pressure across the needle valve in the flow tube changes, causing the flow rate to change. The differential pressure regulator compensates for the change to maintain a constant air flow through the needle valve.

It is recommended that the bubbler tube be kept as small as practical; most designs are calibrated on use of 1/4-in. I.D. tubing. Since air flow rates are low, the best results are obtained when individual air bubbles rather than a stream are seen rising from the end of the tube. Such flow rates are sufficient to exclude water or debris from entering the tube and possibly causing a system malfunction. Tubes larger than 1/4-in. I.D. permit water to rise inside the tube, resulting in the deposit of grease and scum on the inner wall — eventually blocking the flow of air and producing a system malfunction.

Effective purging can be accomplished with small air receivers at 50 psi and constant 1/4-in. diameter tubing. Manual pushbutton purge valves are available. They permit the high pressure air to enter the bubbler tube directly, and at the same time block off the low pressure line to prevent damaging the equipment which regulates the normal bubbler air flow.

Some bubbler systems utilize a pressure responsive transducer instead of switches for pump start-stop signals and alarms. The transducer output is connected to a series of solid state "set point" circuit boards which provide a relay output for initiating the desired function. The set-point cards can be adjusted in the field and usually have LED indicators to show the On state.

These systems have a variety of sophisticated options available, and are subject to the same skilled maintenance requirements as other solid state electronic controls. Solid-state system costs range from the same to higher when compared to electro-mechanical systems. Pneumatic system costs for either type is the same.

The bubbler system is inherently explosion-proof because no electrical equipment is installed in the wet well.

Electrode Systems

Electrode systems are the least prevalent of the liquid level systems. They use a series of specialized relays which rely on small electrical currents (milliamp level) flowing in the liquid from one metallic electrode to another, or to ground. When current flows, the relay is energized; if no liquid is in contact with the electrode, current stops flowing and the relay is de-energized. One relay is required for each function — a pump On-

Electrical

Photo 4-3: Interior view and wiring, electrical control panel. (Photos courtesy: Electric Specialty, Inc.)

Off function requires one relay, alarm level requires another, etc.

Due to the nature of sewage, standard electrodes have not proven satisfactory. A system using a flexible bulb and a clean water solution has been devised, consisting of a 1-1/2 or 2-in. standard pipe, a flexible neoprene bag, and an electrode holder. The neoprene bag is secured to the lower end of the pipe and the electrode holder to the upper end, above the highest water level. Suspended type electrodes are attached to the electrode holder and installed inside the pipe. This assembly is then installed in the wet well and filled with a solution of clean water and bicarbonate of soda (ensuring good conductivity) to the level of the ground electrode.

This system is limited by the volume of the liquid contained in the flexible bag, which holds only a volume of liquid sufficient to fill a 1-1/2-in. pipe to a maximum length of eight feet. Available electrode holders are constructed to NEMA standards but are not designed for the environment of a sewage wet well. Special care is required during installation to ensure that the holder is watertight. However, it is impossible to guarantee that the electrical assembly will remain watertight, especially if submergence occurs during a power outage.

Electrode systems require the same skill level for maintenance as float switch systems, but are higher in cost due to the special relays.

Ultrasonic Level Detectors

Ultrasonic level detection systems are among the most sophisticated liquid level controls. While the same basic components and options offered by other systems are available, a different method is used to determine liquid levels.

Ultrasonic systems use high frequency sound waves. An ultrasonic transmitter is mounted in the wet well and pointed vertically downward. Sound waves strike the water surface and bounce back to the transmitter/detector, which measures the length of time required for the waves to travel.

This time differential is converted by solid state electronic circuitry to a 4-20 ma input signal, which is proportional to the level, to solid state electronic plug-in cards. The input signal can be adjusted to actuate a relay at any desired level. Contacts on the relay are then used to activate the magnetic starters and alarms.

Ultrasonic systems are relatively expensive and require regular maintenance by skilled personnel to insure reliability. In addition, sophisticated test equipment and skilled technicians are required for troubleshooting in the event of malfunction.

Pressure Transducer

Pressure transducers are either mounted in the wet well or connected to a bubbler tube. In both cases, the transducer senses the hydrostatic head of the wet well and converts it to a 4-20 ma signal. The limitations of maintenance and service are the same as described for the ultrasonic system.

Hazardous Area Applications

Municipal wet well lift stations are seldom classified as hazardous locations. However, when they are so classified, the following factors must be considered.

Of the level control systems discussed above, only the bubbler system has inherent intrinsic safety, since it has no electrical wiring or component installation in the wet well itself. The other systems require both electrical wiring and installation within the wet well.

Electrical

The term *intrinsic safety* as applied to electrical or electronic apparatus means simply that any failure of component or wiring cannot produce a discharge of sufficient energy to create a spark or generate enough heat to ignite an explosive atmosphere. The wet well level control equipment must be installed in a non-hazardous area and the level sensing devices must be connected to intrinsic barriers or couplers located in the control panel. In the case of electrode systems, intrinsically safe relays are available which do not require additional intrinsic couplers.

Intrinsic safety as applied to lift stations produces special installation problems. The NEC requires that wiring from an intrinsic device in a non-hazardous area to a sensor in a hazardous area pass through a conduit equipped with a sealed fitting so that there is no passage of hazardous gases into the control panel. A sealed fitting is a standard electrical conduit fitting which is filled with a liquid epoxy after running the wires. The epoxy cures after a brief period to form an impervious barrier to gases; non-hardening compounds are also available.

NEC also requires that power wiring and intrinsic wiring be separated, which for practical purposes means separate conduits for pump power leads and for level controls.

To facilitate maintenance, junction boxes are used to make the connection between the level sensor and wiring from its related intrinsic coupler; a separate intrinsic coupler is used for each float switch except for lead pump On-Off switches, which can utilize a latching type coupler. It is recommended that this junction box *not* be installed in the wet well.

NEMA-rated junction boxes are not designed for a wet well environment with the exception of NEMA 6 (rated as submersible), which itself has limitations as to depth and length of time the box may be submerged. Conduit and sensor connections provide a possible entry of water even under the best of conditions.

NEMA 4 and NEMA 7 junction boxes are sometimes specified — but not recommended — for installation in a sewage wet well.

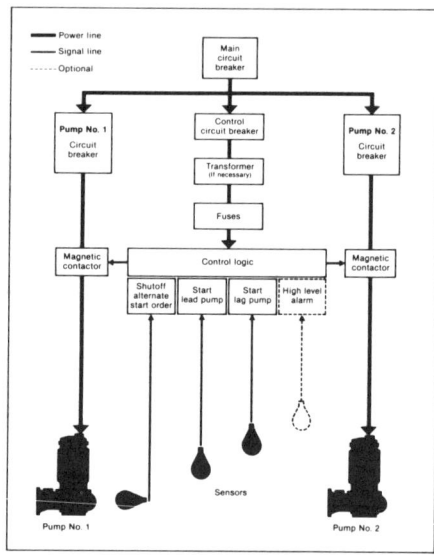

Figure 4-4: Logic block diagram of duplex power panel system. (Courtesy: Flygt, Inc.)

Both of these types can become filled with moisture in a saturated atmosphere and result in system malfunction. If such installations cannot be avoided, it is suggested that the junction box be filled with a waterproof grease after the wires are connected to the terminal strip. Wire-nutted connections wrapped with self-vulcanizing tape are an alternative method. Note that a NEMA 7 junction box is not required with intrinsically safe wiring since, by definition, a wiring fault cannot ignite an explosion.

Operation Sequences

It is common in lift station design to provide a total pumping capacity greater than the anticipated peak flow. In duplex stations, for example, one pump is usually capable of handling maximum anticipated flow, triplex stations may be designed such that no more than two pumps are required to handle maximum flow, etc. This makes it necessary to provide additional circuitry in the control panel to interface with the level sensors for alternating the pumps and starting the back-up pump(s).

Electrical

In duplex stations, the most common technique is to automatically alternate the pumps and provide a lead-lag circuit. The normal sequence is as follows: Water level rises to lead pump On level; the pump starts and lowers the level to pump(s) Off point; the pump stops and automatic alternation indexes to position the second pump as lead pump for the next cycle. If the lead pump starts and the water level continues to rise to the lag pump On level, the second pump will then start and both pumps will operate together until the water level is lowered to the pump(s) Off level. If the lead pump fails for some reason to start, the lag pump is designed to begin operating.

A similar sequence using multi-pump alternators can be utilized for stations with three or more pumps. As the level rises incrementally after the lead pump is started, additional pumps are brought on in sequence until the level is lowered to the Off point and the alternator indexes to the next pump in the sequence.

Alternation may also be accomplished by manual selector switches or time clocks. Time clock alternation is automatic, but instead of alternating pumps at the end of each pump cycle, the alternation occurs at a time of day or day of week as predetermined by the time clock. Manual alternation is accomplished by a selector switch which fixes the lead-lag sequence until the maintenance personnel physically change the switch to select a new sequence.

The purpose of all alternation techniques is to distribute wear evenly on all pumping units and to ensure that a unit does not become inoperable due to idleness.

Auxiliary Equipment

Alarm Systems

Alarm systems are generally required for all lift stations. They may vary from a simple high water level alarm to a multiple system.

For obvious reasons, the most common and necessary is the *high water alarm*. It is recommended that the high water alarm be coincident with the last lag pump in the sequence for multi-pump stations. When a pump station is designed as discussed above, as the level rises to call for a lag pump, failure of the lead pump or pumps is a probable reason. If the alarm is activated by a separate level sensor, it is suggested that the level for activation precede the On level for the last pump.

A *redundant Off level with alarm* has become a popular requirement in applications where the wet well is classified as a hazardous location. The NEC code has often been mistakenly interpreted as requiring explosion-proof motors or pumps installed in such wet wells; in practice, standard submersible units may be used as long as the lowest level maintained is above the top of the pump.

To guarantee submergence in such applications, an additional level sensor is installed to activate slightly below the normal Off level. This level sensor is used in most cases to activate an alarm and to deactivate the pump pilot circuit. The control system must be designed to eliminate the possibility of manually running the pump with the H.O.A. switch, thus ensuring guaranteed submergence.

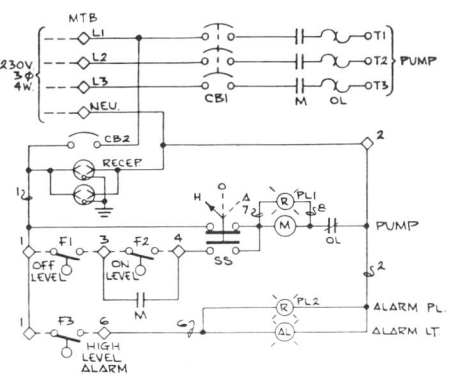

Figure 4-5: Schematic of simplex control system for multi-level 110 volt float control. (Courtesy: Electric Specialty, Inc.)

Electrical

Another valuable accessory is the *pump failure alarm*, which uses as a sensor either a pressure switch installed in the valve chamber on the individual pump discharge just ahead of the check valve, or a limit or tilt switch installed on the check valve lever arm. If the pressure switch or check valve switch fails to open when the related pump is activated, an alarm circuit is energized after a preset time delay and the related pump shut down. Since any pump failure, electrical or mechanical, results in the inability to move the liquid, this alarm provides a first indication of trouble.

A *power failure alarm* may also be required, particularly if the station is equipped for operation by a portable generator in the event of a power failure. A phase monitor is generally used as a sensor in three phase systems and a pneumatic time delay relay in single phase systems. Battery-powered alarm devices are required if local alarms are to be activated. If remote self-powered alarms are used, a dry contact opening or closure can be provided by the power fail sensor.

Seal failure alarms and *motor heat sensor alarms* are sometimes specified. Most manufacturers do not require an audible alarm for warranty purposes. Visual indication of the seal failure condition and inclusion of the motor heat sensor in the pump starter circuit is usually adequate to satisfy warranty requirements.

Other alarms which may be specified are motor overload trip, station flood (failure of check valve to close), low air supply (for bubbler systems), circuit breaker trip, etc. These alarms, with the exception of station flood and low air supply, are of dubious value, since they add to system cost and complexity.

Local alarm devices can be both audible and visual. The preferred alarms are either bell or horn. Visual alarms are usually a flashing light, sometimes with a dim glow feature. These units should be weatherproof since they must be installed in exposed locations if they are to be seen or heard.

Multiple alarm systems usually have an indication light on the control panel and a latching circuit with a Reset-Off-Normal switch for each alarm. The Off position silences the local alarm — remote alarms may have to be reset — and the reset position is used to unlatch the alarm circuit once the condition is corrected. The reset position may also be used to bypass the alarm feature temporarily if the sensor is determined to be at fault.

Telemetry Systems

A variety of telemetry systems are available to communicate an alarm condition to a desired remote location, such as the sewage plant or a police station. These systems range from simple single signal devices to complex *FSK* (frequency shift key) systems. The transmitter portion is usually installed in the control panel and may have its own battery back-up power system. Such devices as tape dialers, digital communicators, tone systems and voice-synthesized computers are available. Tape dialers are generally limited to the transmission of two alarm conditions, whereas FSK systems have the ability to transmit 25 or more separate signals.

Most systems require dry contact inputs from the control panel alarm circuit — either normally open or normally closed, depending upon the particular device used. If only one signal is to be transmitted from a panel with multiple alarms, then a summing relay is added in the control panel to provide a single input to the telemetry system.

Additional Control Components

A completely integrated control panel may also contain a number of other elements required for proper operation of the station. Power distribution panels (circuit breaker load centers) may be included to provide the necessary short circuit protection for additional equipment in or near the station. Panels may include time clocks and contactors to provide automatic cyclic operation of vent blowers. There may be starters and control circuitry for auxiliary equipment.

Electrical

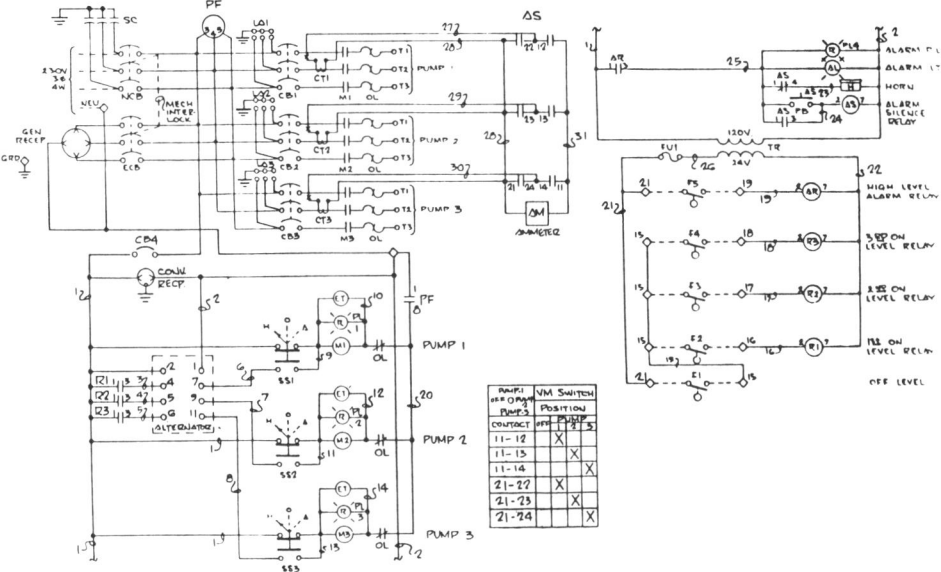

Figure 4-6: Schematic of triplex control panel using a 230 volt system for 12.7 hp pumps. (Courtesy: Electric Specialty, Inc.)

Another device commonly used in control panels installed in exposed locations is the condensation heater. This is usually a 60 to 150 watt unit installed near the bottom of the control panel and controlled by a thermostat set at 60 to 70 degrees F. This heater maintains the air temperature in the panel above the dew point to prevent condensation of moisture and any resulting deterioration of the electrical equipment.

The control manufacturer will help to determine which optional controls are needed for a particular application, and will then fabricate the control panel to this specification. All panels should be built in accordance with NEC requirements.

Control Panel Enclosures

Several manufacturers make a variety of electrical enclosures for installing the type of equipment discussed above. These enclosures are available in a variety of materials (usually steel) and sizes to accommodate any set of conditions likely to be encountered.

Most enclosures meet one or more NEMA ratings indicating suitability for use in specific environments. Some of the more common NEMA ratings are:

NEMA Type 1
General Purpose
To prevent accidental contact with enclosed apparatus. Suitable for application indoors where not exposed to unusual service conditions.

NEMA Type 3
Weatherproof
(Weather Resistant)
Protection against specified weather hazards. Suitable for use outdoors.

NEMA Type 3R
Raintight
Protects against entrance of water from a beating rain. Suitable for general outdoor application not requiring sleetproof.

Electrical

NEMA Type 4 **Watertight**	Designed to exclude water applied in form of hose stream. To prevent against stream of water during cleaning operations, etc.
NEMA Type 4X **Corrosion Resistant**	Designed to exclude water supplied in form of hose stream and used in areas where a serious corrosion problem exists.
NEMA Type 6 **Submersible**	Intended to permit enclosed apparatus to be operated successfully when submerged in water under specified pressure and time.
NEMA Type 7 **Hazardous Locations** **Class I — Air Break**	Designed to meet application requirements of National Electrical Code for Class I, Hazardous Locations (explosive atmospheres). Circuit interruption occurs in air.
NEMA Type 12 **Industrial Use**	For use in those industries where it is desired to exclude dust, lint, fibers and flyings, or oil or coolant seepage.

All of these enclosures may be fitted with an inner door for mounting pilot lights, switches, gauges, elapsed time meters, and other components. With an inner door unit, no live parts are exposed and there is a barrier between the operator and all live components. This type of construction is also desirable in outdoor applications to discourage vandalism.

Grinder Pump Panels

Most grinder pumps are used in low-pressure sewer systems for one or more single-family dwellings. They are simplex units with single-phase motors. These motors may require start and/or run capacitors, plus a control relay that must be installed in the control panel. However, if only run capacitors are used, a control relay is not necessary. Overload relays for motor overload protection are required.

Motor heat sensor cut-out (a thermal switch in motor winding) and seal-fail indication options are offered by most manufacturers. Specific component additions to the control system are needed to utilize the optional equipment properly.

Some grinder pump manufacturers offer a delay timer option so that the pump run cycle can be extended below the Off float level and permit the pump inlet to break suction. This is done both to clear any floating solids from the liquid surface and to ensure that no solid — such as wood, plastic or other difficult-to-chop material — is left in the primary cutting portion of the pump to cause jamming on the next cycle.

Reversing contactors have also been used for the same purpose. In this design, the direction of rotation of the pump is reversed at the end of each cycle to clear the primary cutters of solids.

However, because of added costs, neither the time delay nor the reversing options are commonly specified today.

Many simplex grinder pump installations, while serving individual dwellings, are municipally owned and maintained. This means putting the control panel on the exterior of the dwelling, which necessitates weather-proof enclosures (NEMA 3R, 4 or 4X styles).

Larger grinder pumps are available with 3 and 5 hp single phase motors and 3, 5 and 7½ hp three phase motors. These pumps are most commonly used in duplex lift station applications. The same control parameters apply as presented in this chapter for single and three phase control panels and accessories.

Chapter 5
Mechanical Controls and Components

Properly specified and installed mechanical controls and components are important for the easy operation and maintenance of a submersible wastewater lift station. These controls and components include valving, for hydraulic control; access doors, for easy entry; and slide rail assemblies, for convenient pump disconnect, removal, and re-insertion.

Wastewater Valves

Valves used in pipelines where wastewater is pumped differ from those used on water lines because more consideration must be given to the amount and character of the solid matter in the wastewater. For example, there is a tendency for fibrous materials to lodge and build up on the internal parts of a valve — and particularly on the orifice seat. This can result in clogged valving, causing serious operating and maintenance problems within the entire system.

As a result, it is recommended that valves be used only when they are essential to the operation of the lift station, and that they be so designed and located that they are either fully opened or fully closed at all times. Valves which permit the throttling of liquid flow have no place in sewage system design.

Both a check valve and a shut-off valve should be installed on the discharge side of each pump. They should be installed even if not required by code authorities. The purpose of the check valve is to prevent backflow when the pump or pumps reach the Off position, or in the case of failure of the power supply to the pump. Use of check valves in multiple pump installations eliminate reverse flow through the non-operating pump or pumps. The shut-off valve is essential to stop system flow during pump or check valve servicing.

All valves should be installed outside of the wet well, in a separate box or vault. The valve controls are then easily accessible for servicing and, under normal conditions, there is no need to send personnel into the wet well.

Check Valves

Swing Disc. A typical check valve is a swing disc (**Figure 5-1**), which is preferred over the tilting disc type for sewage applications. Swing check valves can be provided with either an outside spring and lever, or an outside weight and lever. They are normally installed in a horizontal position to minimize the buildup of grit and solid matter on the internal parts of the valve. Some swing check valves have an access port to

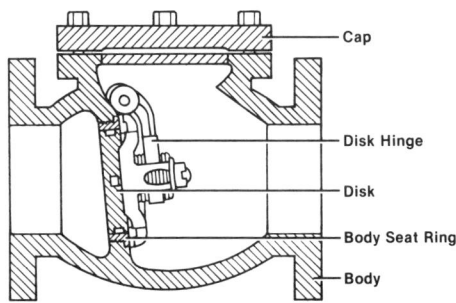

Figure 5-1: Swing disc check valve.

Mechanical

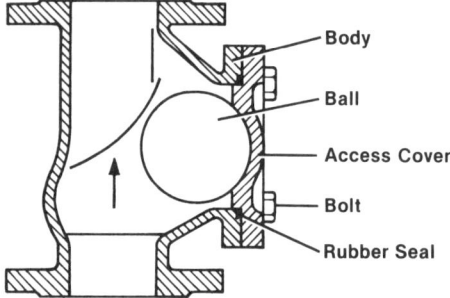

Figure 5-2: Ball check valve.

remove the disc for cleaning or replacement without removal of valve from line.

Ball Check. The recent development and expanding use of ball check valves (**Figure 5-2**) has eliminated some of the problems commonly found with a traditional swing check valve in submersible wastewater pumping. These valves use a corrosion-resistant ball rather than a hinged disc. The ball continuously spins by virtue of the fluid flow in the valve body and has a tendency to clean itself. Therefore, a ball check valve can be used in either a horizontal or vertical position.

An access port is commonly furnished to allow for easy inspection of the valve without removing it from line. Proximity switches can be mounted in some ball check valves for monitoring mode of operation. A suitable back pressure should be available for proper sealing of the valve.

Shut-off Valves

The most commonly used shut-off valves in wastewater pumping systems are the following:

Plug Valves. A typical plug valve is shown in **Figure 5-3**. They are commonly made with a cast iron body and seat construction. Most plug valves provide a valve port which is equal to the pipe size, to prevent any restriction of the flow path. The valve is fully opened or closed by simply turning the valve stem 90 degrees, or a 1/4 turn. In larger installations, these valves are available with gear actuators or motor-driven devices. They should be opened and closed periodically to assure ease of operation.

Plug valves are popular for wastewater control. While they are relatively expensive for smaller applications, they are preferred for larger stations, in remote locations, and in cases of low flow, since they allow only a minimal accumulation of grit and other solids on the valve seat.

Ball Valves are increasingly popular. They are very similar in construction to plug valves; both provide full flow opening and open or close with a 1/4 turn of the handle. Ball valves, in general, provide a better seal than plug valves and are equally easy to operate. (**Figure 5-4**)

Gate Valves are commonly used in submersible lift stations, because they provide both a full flow opening and it is easy to extend the stem upwards to ground level for simple manual opening and closing. There is a tendency, however, for solids to accumulate on the valve seat. Normal construction of gate valves is either bronze or, for larger sizes, cast iron bodies with bronze trim.

Gate valves (**Figure 5-5**) are commonly available in both rising stem and non-rising stem designs. Rising stem types are preferred because they make it easy to see whether the valve is open (the stem rises) or closed (it goes back down). The stem

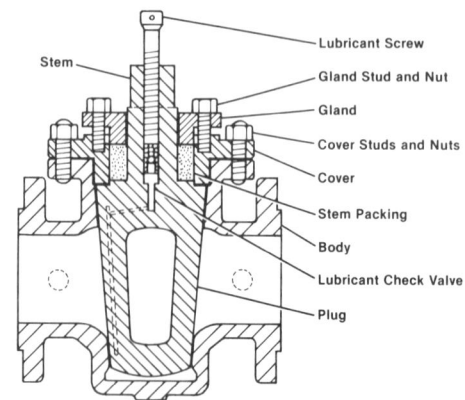

Figure 5-3: Plug valve.

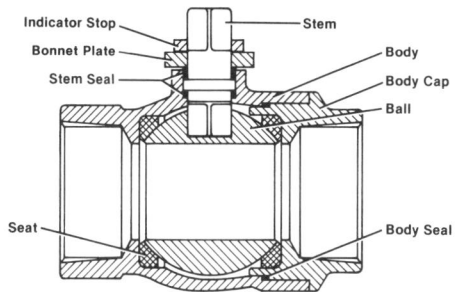

Figure 5-4: Ball valve.

threads up into the bonnet when the valve is open, keeping it out of the flow path and reducing the chance of solids accumulation and clogging. However, non-rising stem types are often specified when there is a clearance problem, because the stem remains in the same location whether the valve is open or closed.

Electric actuators are available to open or close most standard gate valves, particularly in larger installations. Motor-driven valves are particularly suited for automatic or remote control.

Water Hammer

Water hammer occurs when liquid flowing in a pipe is abruptly stopped, causing an energy release which must be absorbed between the fluid itself and the piping system. Water hammer is a series of shocks sounding like hammer blows, which may have sufficient magnitude to rupture the pipe or damage the check valve.

When designing a submersible wastewater lift station, it is important to protect the piping system from damage by water hammer. This can be addressed by considering the use of a power-operated valve that will close, or start to close, before pump shut-off. In a long run of pipe, it is often advisable to install several check valves spaced well apart, or to use a surge valve to compensate for the excessive pressure which is created when a pump is shut off or a valve is closed. The surge valve is typically installed parallel to the pump and the discharge valve, with a drainback line from the surge valve to the lift station itself.

Mechanical

Strainers

Trash baskets (strainers) are often specified where there is the possibility of objects larger than the pump can handle entering the wet well. Baskets avoid clogging of the pump and permit easy removal of such objects. They may be installed ahead of the station or at the well inlet. Provision must be made for removing and cleaning the basket without entry into the wet well.

Maintenance

It is important that all piping and valves be inspected regularly to ensure that unnecessary loads are not imposed on the submersible pump. Additional head loss is created by improper maintenance of the discharge pipe of the pump, which can increase the wear rate and decrease the efficiency of the pump. It is advisable to check discharge pressure routinely, to ensure that the pump, the piping, and the valves are not partially clogged or built up with greasy material. When stem-operated valves are used, they should be lubricated regularly according to the manufacturer's instruc-

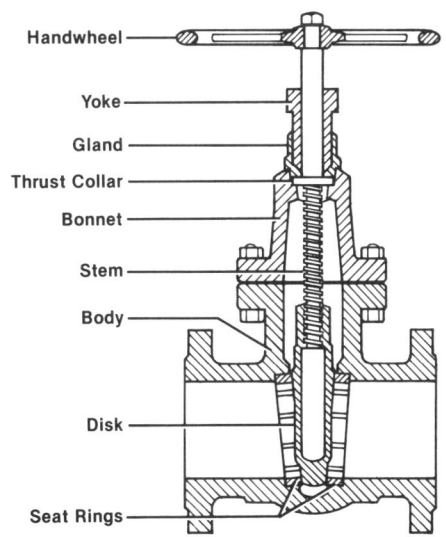

Figure 5-5: Gate valve.

Mechanical

tions and the stems rotated periodically to ensure ease of operation.

If an obstruction is caused by a buildup on the inside of a valve, it may be possible to back-flush and eliminate it by using high pressure water. If this fails, entry to remove the buildup can be made through the valve access port or it may be necessary to dismantle the pipe at the point of constriction.

If the gradual buildup of grease and materials is permitted in the piping and valves, serious problems may occur. This buildup may be, in many cases, difficult to dislodge either mechanically or with chemicals. If it is a continuing problem, there should be a routine procedure to anticipate and remove it before the problem becomes severe.

Access Covers

Access covers are used in submersible lift stations to cover and provide access to the wet well and, when used, to valve boxes.

The simplest access cover, usually installed on a very small station, is a manhole cover. It is round rather than square or rectangular, to avoid having it fall into the station. It also has no hinges, and must be lifted manually for personnel to gain access to the station.

Easier, safer access is provided by fabricated covers with hinged doors. When closed, these doors can have locking provisions to limit access to authorized personnel only. When hinged doors are opened to the vertical position, they provide full access to the station and to the valve box. (**Photo 5-6**)

The hinged cover can be fabricated of cast iron, steel or aluminum. It is available with or without a spring assist. The door must be capable of being locked closed and, preferably, should be able to be locked open.

In some applications, submersible lift station access covers must be weathertight. This is usually important when metered sewage, not mixed with stormwater, is processed through the station. The weathertight access cover is designed with a door leaf or leaves overlapping a gutter or channel type frame. The surface water is then drained through the frame and away from the station. This design can also be fitted

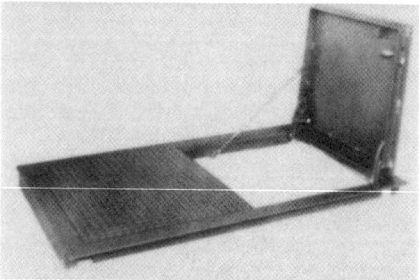

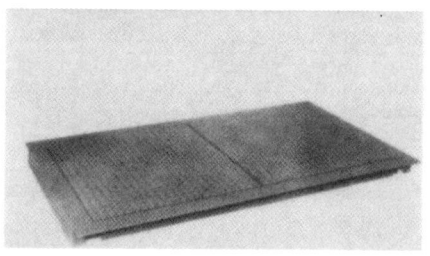

Photo 5-6: Access door for submersible station. Duplex well access door covers — hinged fully open (top), partly open, and closed. (Courtesy: The Bilco Company)

with a gasket or other membrane to reduce the passage of sewer odors emanating from the station. To make the membrane effective, the installer must carefully set the frame and shim to ensure that the doors close evenly all around the frame and gasketing.

If a lift station must be located in an area subject to flooding — or in a location where the access cover must be gas-tight — the construction of the unit would be the same. That is, it must be suitably designed with a means of compressing a gasket between machined surfaces on the cover and frame

Mechanical

all around the perimeter of the frame opening. This is done with lever type dogs, or bolt-down devices, on all four sides and on corners if necessary. This type of door is high in cost and available through a limited number of manufacturers.

The size of the access cover for the wet well should be coordinated with the station design, the number and size of pumps, and the guide rail removal system. The main purpose is to allow adequate free space for installation and removal of pumps and, if necessary, entry into the wet well.

Nominal sizes are usually given as frame opening sizes. Free space may be somewhat less due to the door stop, open door latch, reinforcing members, and handle mechanism.

An important factor in access cover selection is the amount of reinforcement, or load capacity. The standard access door is designed to carry either 150 or 300 lbs. live load per square foot. While 150 lbs/psf is sufficient to carry pedestrian traffic, many codes require 300 lbs/psf units; these are available but are not necessary for normal pedestrian traffic.

For heavier loads, both fabricated and cast *heavy duty access covers* are available. These units are able to handle 16,000-lb. (HS-20) wheel loading in alleys, parking lots, and other off-the-road traffic areas. Fabricated units should *not* be installed in streets, highways, or other areas subject to high-density traffic. In these applications, it is recommended that heavy-duty cast iron or cast aluminum manhole covers be specified.

Aluminum is the most widely used cover material, since it offers adequate strength and provides for low maintenance in corrosive damp or wet conditions. Aluminum units are available with an anodized finish. Steel access covers may be furnished with hot dip galvanized, coal tar epoxy or other properly prepared corrosion-resistant finishes.

Specifiers also have a choice of single, double or triple door designs. If the station is of a reasonable size, a single door may be used to cover the entire opening at lower cost than a double or triple door. Double door units are usually recommended for duplex and larger pump stations.

Fabricated access cover doors are normally constructed of 3/16 or 1/4-in. diamond pattern plate or other non-skid surface of aluminum or steel. Stiffening members are welded to the underside for proper reinforcing.

Frames are normally constructed of 1/4-in. aluminum or steel. One common construction is a 3 × 3-in. or 3 × 2-in. angle or aluminum extrusion. For a weathertight installation, the common construction is a channel-shaped frame.

Hinges are either cast aluminum or steel, normally cadmium or zinc-plated. They may also be forged or formed stainless steel, or forged or formed brass. The hinges should be fastened to the cover with tamper-proof lock nuts and bolts. All other hardware items are also available in stainless steel.

There can be a provision on the access cover frame for bolting the upper guide rail brackets, either in a fixed or adjustable position, depending upon the guide rail design. Provision can also be made for mounting cable holders and chain hooks.

Guide Rail Systems

An essential component of a submersible wet well system is a method of connecting and disconnecting the pump from discharge piping within the wet well.

Guide rail systems permit installation and removal of the pump unit from the top of the wet well. This ends the dirty, time-consuming and potentially dangerous task of sending service personnel into the station. **Figure 5-7** illustrates how one system works.

The wet well floor is generally designed to accommodate submersible pumps in such a way that solids will not settle out in "dead" spots and become septic, creating a potential odor problem. Therefore, the pump should be guided to a specific position in the sump where it will engage with the discharge piping.

The guide rail or rails are usually made of

Mechanical

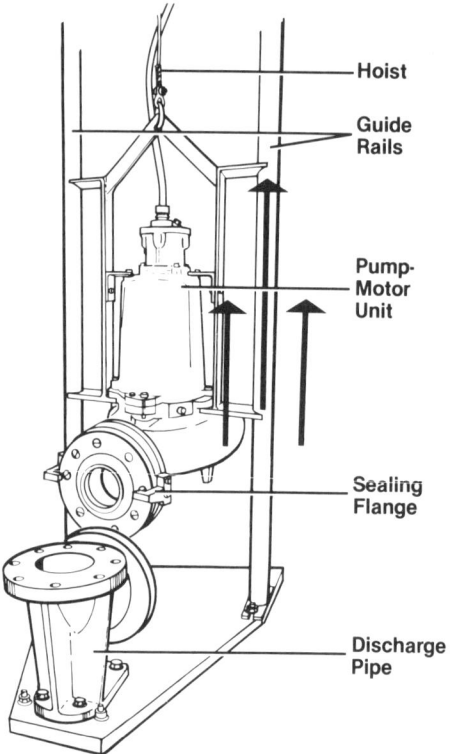

Figure 5-7: How one style of guide rail system works. The pump-motor unit is disconnected, lifted from the station, and dropped back into place. Sealing is by means of a quick-disconnect flange. (Courtesy: Hydromatic Pumps)

pipe — or sometimes angle, channel or cable — to accommodate the pump manufacturer's guide brackets that are attached to the pump. The rails should be of rigid material with suitable metallurgy to withstand the action of liquids and vapors for a reasonable life expectancy of the pump station. Depending on the depth of the wet well, one or more intermediate guide rail supports may be required to maintain sufficient rigidity for smooth guidance of the pump unit.

The location of the upper guide rail brackets is important. While most submersible pumps are designed with the rail guides on the front of the pump casing, others have one rail on either side. With the latter, provision must be made across the access opening to connect the upper guide rail bracket between the pumps in a duplex or multiple pump station. The access door manufacturer can furnish integral provisions on the door frame to accommodate these brackets — or they can be field-installed by the contractor as part of the well cover structure.

The lifting or lowering of the pump unit should be done with either a chain or cable securely attached to the pump. It should be compatible with the hoist or lifting device used, and should have a minimum 4 to 1 safety factor over the weight of the pump and guide brackets.

The actual seal between stationary and movable components may be accomplished in varying ways, depending on the individual pump manufacturer. In all cases, the manufacturer should be able to assure the owner of a good connection that is free from excessive leakage. All threaded and flanged connections between guide rail components and other piping must be in accordance with ANSI standards.

Chapter 6
Installation and Start-Up

Proper installation of submersible wet wells involves careful coordination from planning through completion. Deliveries must be scheduled and checked. All accessories must be installed in accordance with the job specifications and manufacturers' instructions. Electrical control panels must be installed and tested with care. When start-up is accomplished properly, the specifier, operator and owner can be assured of trouble-free operation.

Careful preparation and planning is needed to ensure proper installation of all equipment in submersible pumping stations. There must be effective coordination between the supervising engineer, the mechanical and electrical contractors, and all suppliers.

The first step is to make certain that the required station equipment is ordered completely and accurately according to specifications, and that — considering lead times — all of it will arrive prior to the scheduled installation. This includes not only the pump and electrical control panel, but the station itself (if prefabricated) and all accessories, such as piping, valves, and access covers.

Receipt and Inspection

All equipment should be examined upon receipt for any signs of apparent damage. If damage is indicated, a claim should be filed immediately with the carrier and the supplier should be notified. All parts shipped loose or separately should be checked for loss or damage.

If the equipment won't be installed immediately, it is best to store it in a clean, dry location where it will be protected from possible damage. When storing equipment, be sure to follow the manufacturer's recommendations. Check it prior to storage to anticipate any possible problems at the time of installation. By checking sizes, design features called for on the plans and specifications, and all interface components, installation problems can be minimized.

Handling Pumps

A submersible pump should be transported and stored in accordance with the pump manufacturer's recommendations. Make sure that it cannot roll or fall over. Always lift the pump by its carrying handle, never by the motor cable.

In the absence of specific manufacturer recommendations about storage, pumps must be protected against excessive moisture and heat. This precaution is necessary to prevent the possibility of moisture damage to internal components and the power cord, etc. The impeller should be rotated by hand occasionally (for example, every other month) to assure free movement of the rotating elements. Make sure that the cable entry seal conforms to the outside diameter of the cable to prevent leakage into the pump.

The end of the power cable must *not* be submerged, as water may wick through the cable into the motor.

After any period of storage, the pump should be inspected, tested and/or reconditioned in accordance with the manufacturer's recommendations before it is put into operation.

Installation

Pump Station Installation

When beginning the installation, make certain that all equipment is on site and that all manufacturer instructional literature is available and has been reviewed.

In addition to the pump(s), the following items are required:

- Discharge elbow/support base and/or sealing arrangement for connecting the pump to the discharge line.
- Guide rail(s) consisting of specified material and size.
- Upper guide rail brackets for attaching the rail(s) to the access cover frame or top of the station.
- Intermediate guide rail supports as required.
- Specified discharge piping and fittings.
- The proper check valves and shut-off valves.
- Level sensor or other control equipment.
- Cable holder for level sensors or other sensor brackets.
- Junction box or conduit box (if required).
- Access frame (with covers).
- Control panel.

Access Cover Installation

There are two common installation methods for access covers. One is to cast the unit into the poured concrete slab well cover; remember that aluminum frames must be protected from the wet concrete by a bituminous coating. The other method is to frame flange the unit and drop it into the steel well cover.

Both types of installations must align properly with the base plate. Prior to casting the door unit into the concrete, the cover should be closed and checked to make certain that it rests on the frame all around. Shim the frame as necessary to ensure proper door closure.

Installation procedures when casting the access cover in concrete are as follows:

- Spring-loaded units — *Caution: cover is spring loaded.* Do not remove safety shipping bolt until unit is ready to be installed in a normal horizontal operating position.
- Non-spring loaded units — Shipping bolt or banding should not be removed until the unit is ready to be installed and is in normal horizontal operating position.
- On angle frame type doors, be sure concrete anchors are in proper position when setting frame. Then remove banding or safety shipping bolt and plate, and outside latch release handle.
- Place the access frame and cover on the opening so that the lift handle is in desired location and the slide rail connections, junction box or conduit box provisions, and chain hooks (if all are furnished) are in proper relationship to the pump base plate and the discharge piping connections.
- Before pouring concrete or anchoring in place, open and close door(s) and check to see that the door rests on frame all around. If not, shim under frame as needed at corners and recheck for proper door closure.
- With door(s) closed, place the access frame in position and install the anchor bolt. Pour the concrete flush with the top of the frame. Use care to prevent the concrete from getting into the frame or around hinges. Be sure to support the frame to prevent sagging. Do not permit the weight of the concrete to push frame inward and reduce the clearance between door(s) and frame.

There is little or no maintenance required on access covers. Painted doors should be cleaned and repainted when necessary. Spare parts — such as springs, hinges, latches and arms — seldom require replacing, but should be readily available from the supplier. To ensure that replacement parts are ordered correctly, the broken part should be returned to be matched. If this is not possible, the model number, size, type of metal, year of purchase and shop order number should be provided.

Internal Assembly

Now that the access cover is in the proper position, you can begin to install the internal station components.

Place the pump discharge connection in

Installation

position. Temporarily secure the guide rail(s) in the upper mounting brackets and the discharge bosses at bottom. Install the intermediate support brackets, if required. Make sure the rails are in a true vertical position, so the pump will clear the access opening and will slide freely down the rails into place in the discharge connection.

Once the rails are in proper alignment, bolt the discharge connection into the floor of the station. Connect the discharge pipe to the discharge connection and proceed to install the check valve, shutoff valve and fittings according to the plans and specifications.

All level-sensing devices must be properly secured to the access cover frame or the wet well. Install them at the levels indicated on the plans and specifications. Control components, other than level-sensing devices, should not be installed in the wet well.

Lifting equipment is normally required for handling pumps. It should be able to hoist a pump straight up and down in the station, preferably without having to reset the lifting hook. Make certain the lifting equipment is securely anchored. Keep all personnel out from under suspended loads.

Before lowering a pump into place, it must be checked for correct rotation, using one of the following methods:

1. Prior to installation in the wet well, lay the pump on its side and momentarily run (jog) and check it for rotation. When running the pump outside the wet well, care must be taken to stay clear of the impeller and to provide a safe temporary connection of the motor leads. The best procedure is to lift the pump with a hoist or tilt it slightly and observe the rotation. An arrow on the pump casting, or instructions in the operation and maintenance manual, will show the direction for proper rotation.

This is the most accurate method to ensure proper rotation; however, other methods are usually provided as a secondary check or for circumstances when the pump cannot be lifted from the station.

2. After installation, if the check valve has an external operating arm, observation will indicate which rotation of the motor opens the check valve widest. This indicates the higher flows, and is the correct rotation for the pump.

3. If pressure taps are available, rotation can be ascertained by reading the shut-off head on a pressure gauge. The proper rotation will produce the higher pressure.

4. The least desirable procedure is to position the pump in the Hand mode, using the H.O.A. switch, lower it to the bottom of the wet well, and check the drop in water level in the well. The rotation resulting in the faster drawdown is the proper rotation.

One of the above procedures must be used to check and ensure proper rotation of each pump before start-up of the station. Keep in mind that visual inspection of the rotating elements is the most accurate method.

The direction of rotation on three-phase pumps may be changed by interchanging any two motor leads at their control panel connections. For single phase pumps, if improper rotation is observed, consult the operation and maintenance manual or contact the pump manufacturer.

After the proper rotation is verified, lower the pump along the guide rail(s). Upon reaching its bottom position, it should automatically connect to the discharge. Adjust as required, following the manufacturer's recommendations.

Fasten the lifting chain on the access frame eyebolt and fasten the cables on the cable holder. Cable supports are required for deep installations. Run the pump and level control cables up to the electric control panel or into the junction box, if used.

The pump and level control cables should be long enough to reach the control panel without splicing. If a junction box, conduit box or quick connector plugs are utilized, they should be located outside the wet well — or be of at least NEMA 4 construction if located within the wet well.

Arrange for a cable if needed between the sump and the electrical control panel. Make sure the cables are not sharply bent or pinched, and that all connections are sealed and watertight to prevent leakage from ground water.

Installation

Control Panels

Before making a new installation, a qualified electrician or factory service technician should verify the horsepower, voltage rating, and full load amperage of each pump. This information should be used to ensure that the control panel is of the correct horsepower, and that the heater coils furnished in each motor starter overload unit are sized or set correctly to match the motor's operating current as given on the motor nameplate.

The service voltage and frequency should be checked to ensure that it is the same as the motor rating. This information should be used to cross-check the circuit breakers, fuse or disconnect ratings. All electrical work must comply with national and local codes and regulations.

At the time of installation, the panel or equipment should be checked for missing or loose components — including a correct wiring diagram. All wire terminations should be checked for tightness. Care must be taken in handling the control panel and equipment during installation to avoid damage to the enclosure or any of its components.

Location and provisions for mounting the panel should be shown on the plans for the job. Make certain the enclosure is of the type specified for that particular location. Adequate racks, mounting brackets, and fastening hardware must be used to securely mount the panels and devices.

On flat surfaces, panels should have a slight clearance between the back of panel and the mounting surface to allow for circulation of air. While clearance is usually provided by the enclosure design, it may need to be furnished by mounting hardware. This assures heat dissipation and prevents moisture accumulation.

If the panel is mounted on a rack, it must be constructed and braced to provide a flat, rigid surface that will not distort the panel and cause possible door alignment problems.

Mounting bolts and hardware must be of sufficient size to provide stable positioning of the panel.

Field Wiring

All conduits and wires must be installed as required on the plans for the specific job. All cable entrances into the enclosure must be in accordance with NEC, and shall maintain the integrity of the NEMA enclosure, to prevent intrusion of moisture, dust, and gas vapors. When possible, conduits should be attached from the bottom of the enclosure to facilitate weatherproofing. Weatherproof hubs and sealed fittings should be used.

The size of conduits and wires must be adequate for the specific requirements of the incoming service, pump motor leads and remote control devices. NEC must be used as a minimum guideline.

After the installation of the conduit and wiring, make certain that all terminations in the control panel conform to the panel manufacturer's diagrams and instructions. The incoming service voltage must be correct for the panel. Typically, three-phase systems are designed for a 208, 240, 480, or 575 volt, four-wire service voltage. Single-phase systems are usually 208 or 240 volt, two- or three-wire, but are sometimes 120 volt, two-wire service voltage.

As a safety precaution, prior to connecting the motor leads to the control panel or applying power to the pump, megohm readings should be taken and recorded with a 500-volt megger. Connect all motor leads together and check the combination to ground. The readings should be above 20 megohms at all points. A motor should not be run if any reading is below 10 megohms. If this is the case, the source of the low readings should be found and corrected. If the readings are between 10 and 20 megohms, the pump should be run for short periods and the readings rechecked. Only after tests are complete should the motor leads be connected permanently to the control panel.

The pump leads must now be connected to the control panel following the panel diagram, using the pump manufacturer's wiring diagram to distinguish and verify the connection and color coding of the leads. Internal pump/motor safety controls should be identified following the pump manufac-

Installation

turer's wiring diagram and connected to the panel as specified by the panel manufacturer. Make certain the pump is correctly grounded.

If automatic control of the pumps is provided by liquid level sensors in the wet well, all connections must be made to the proper terminal points in the control panel. Care must be taken to identify each sensor and its specific function. Proper connections ensure sequential and automatic operation of the pumps.

Wires must be marked to provide future identification should re-connection be required during maintenance or troubleshooting. A record should be made in the permanent file for the control panel regarding field wire connections, wire sizes and types, cable lengths, and any other related information. This file can be valuable during normal maintenance, for emergency situations, and as a reference for future installations.

When incoming service voltage is available and the system is complete and ready to be put into service, a qualified electrician should be present. If the service voltage is 240 volt, three-phase, four-wire from Delta-connected transformers, the "high leg" must be identified; it has a higher than 120 volt (usually approximately 210 volt) reading phase to neutral. The control panel must then be checked to make certain no control circuits or 120 volt operating devices are connected to this incoming high leg.

Some local codes require that the high leg be on a particular phase connection, usually Phase B or sometimes Phase C. Most control panels are designed to use Phase A for all 120 volt phase to neutral circuits when no control transformer is used.

Start-Up and Testing

Many pump and panel manufacturers have special forms that can be used during the start-up of the station. **(See Figure 6-1.)** Some manufacturers require that these forms be filled out and returned to ensure warranty on the pumps, control panel, and station components. This type of form provides a detailed description of the procedures and tests to be performed. The following summarizes these procedures.

The panel is ready to be put into service after the incoming service voltage has been checked on the line side of the main circuit breaker or disconnect, and it is verified that all phases and neutral (if used) are present and at acceptable levels.

All circuit breakers and selector switches must be in the Off position. The main circuit breaker or disconnect can then be turned On. One at a time, each of the other circuit breakers should be turned On, to check the load side of each for correct voltage. This includes the control circuit.

With sufficient water in the well, turn each pump operating selector switch (H.O.A.) to the Hand position to run each pump. Amperage readings should be taken and recorded on each motor lead with a clamp-on ammeter. The phase-to-phase voltage must be checked and recorded at this time. If there is excessive amperage draw on one leg, start troubleshooting by checking the manufacturer's recommendations. Consult with the power company only after all other checks have been made.

The initial readings for amperage, voltage and ohmmeter resistance, plus megohm readings, should be the start of a permanent maintenance file. Monthly, quarterly, or annual readings should be taken as part of a good preventive maintenance program. They are the basis for scheduled checks, which can indicate the trend of the motor current draw and can help prevent major outages and costly motor rewind jobs.

Operational Checks

When start-up and testing is completed, the system is ready to be checked for automatic operation. Both or all of the pump operating selector switches should be turned to the Auto position, and the alarms turned On or reset as required for normal operation.

The best way to check and set correctly the On and Off levels is to provide an external water supply source to fill the wet well to various predetermined levels. This water can be from a nearby hydrant or other

Figure 6-1

START-UP REPORT

This report is designed to insure the customer that customer service and a quality product are the number one priority.

Please answer the following questions completely and as accurately as possible. Please mail this form to:

(Manufacturer's
Name and Address)

1) Pump Owner's Name _____
   Address _____
   Location of Installation _____
   Person in Charge _____ Phone _____
   Purchased From _____

2) Model _____ Serial No. _____
   Voltage _____ Phase _____ Hertz _____ Horsepower _____
   Rotation: Direction of Impeller Rotation (Use C/W for clockwise, CC/W for counter-clockwise) _____
   Method Used to Check Rotation (viewed from bottom) _____
   Does Impeller Turn Freely by Hand _____ Yes _____ No

3) Condition of Equipment _____ Good _____ Fair _____ Poor
   Condition of Cable Jacket _____ Good _____ Fair _____ Poor
   Resistance of Cable and Pump Motor (measured at pump control)
   Red-Black _____ Ohms Red-White _____ Ohms White-Black _____ Ohms
   Resistance of Ground Circuit Between Control Panel and Outside of Pump
   _____ Ohms
   *MEG Ohm Check of Insulation:
   Red to Ground _____ White to Ground _____ Black to Ground _____

4) Condition of Equipment at Start-Up: Dry _____ Wet _____ Muddy _____
   Was Equipment Stored: _____ Length of Storage: _____
   Describe Station Layout _____

5) Liquid Being Pumped _____
   Debris in Bottom of Station? _____
   Was Debris Removed in Your Presence? _____
   Are Guide Rails exactly Vertical? _____
   Is Base Elbow Installed Level? _____

6) Liquid Level Controls: Model _____
   Is Control Installed Away from Turbulence _____
   Operation Check:
   Tip lowest float (stop float) all, pumps should remain off.
   Tip second float (and stop float), one pump comes on.
   Tip third float (and stop float), both pumps on (alarm on simplex).
   Tip fourth float (and stop float), high level alarm on (omit on simplex).
   If not our level controls, describe type of controls _____
   Does liquid level ever drop below volute top? _____

7) Control Panel Model No. _____
   Number of Pumps Operated by Control Panel _____
   *NOTE: At no time should hole be made in top of control panel, unless proper sealing devices are utilized.
   Control Panel Manufactured By Others: _____
   Company Name _____
   Model No. _____
   Short Circuit Protection_____ Type___
   Number and Size of Short Circuit Device(s) _____ Amp Rating _____
   Overload Type_____ Size_____ Amp Rating_____
   Do Protective Devices Comply With Pump Motor Amp Rating _____
   Are All Connections Tight? _____
   Is the Interior of the Panel Dry?_____. If "No," the Moisture Problem Must Be Corrected.

8) Electrical Readings:
   Single Phase:
   Voltage Supply at Panel Line Connection, Pump Off, L1, L2 _____
   Voltage Supply at Panel Line Connection, Pump On, L1, L2 _____
   Amperage: Load Connection, Pump On, L1 _____ L2 _____
   Three Phase:
   Voltage Supply at Panel Line Connection, Pump Off, L1-L2____ L2-L3____ L3-L1____
   Voltage Supply at Panel Line Connection, Pump On, L1-L2____ L2-L3____ L3-L1____
   Amperage, Load Connection, Pump On, L1 _____ L2 _____ L3 _____

9) Final Check:
   Is Pump Seated on Discharge Properly? _____ Check for Leaks _____
   Does Check Valves Operate Properly?_____
   Flow; Does Station Appear to Operate at Proper Rate _____
   Noise Level: High_____ Medium_____ Low_____
   Comments: _____

10) Equipment Difficulties During Start-Up: _____
    _____
    _____

11) Manuals:
    Has Operator Received Pump Instruction and Parts Manual? _____
    Has Operator Received Electrical Control Panel Diagram? _____
    Has Operator Been Briefed On Warranty? _____
    Address of Local Representative/Distributor_____
    _____

12) I Have Received the Above Information _____
                                                            Name of Operator

    _____
        Name of Company   Date
    I Certify This Report To Be Accurate _____
                                                              Name of Start-Up Man

    _____
        Employed By   Date
    Date and Time of Start-Up _____
    Present at Start-Up:
    ( ) Engineer _____ ( ) Operator _____
    ( ) Contractor _____ ( ) Other _____

Installation

source. Care should be taken when discharging water into the wet well not to affect floats or other controls. By controlling the fill and then observing or setting the operating points, the automatic cycle or cycles of operation can be checked, no matter what type of level controls are employed.

To achieve second and/or subsequent pump start levels, the circuit breaker or disconnect can be turned Off for the lead pump. This allows the wet well to fill, without running the lead pump, to the lag pump On level for checkout.

All pumps may be turned Off to allow checking and testing of alarm levels. The pumps should be permitted to pump, to allow checking of the desired Off level. If a low alarm level is included below the normal Off level, a pump may have to be run in the Hand mode to check it. If guaranteed submergence with redundant low water cutoff is specified, it should be checked to be sure the wiring has eliminated the possibility of Hand operation.

The control sytem should be cycled more than once to check the proper and automatic alternation sequence of the pumps. If the panel is supplied with a manual alternator switch, it should be operated and the pumps cycled again to check for sequencing. Each run cycle, after alternation, should be checked to run both lead, lag and any subsequent pumps to ensure complete, correct alternation of the duty cycle.

Simulation Testing

The following steps should be taken only in those extreme cases where insufficient water is available and the pump(s) cannot be run in the well. The validity of this check-out procedure is questionable, since the pumps are not run and the actual well water level is not used to energize the level sensing devices. It is *not* recommended except as a last resort.

If automatic controls are used with level sensor switches, they can be removed from the wet well and the controls manually actuated. Care must be taken to mark each sensor. With all controls in the Off position, a low wet well level is simulated. By actuating the controls (either manually or by lowering them into a large container of water) one at a time, a rising wet well level is simulated.

It is very important that the correct sequence is achieved when simulating a rising level and also as the floats are turned back to a vertical position or raised out of the container of water. When simulating automatic operation, enough water must be maintained in the wet well to allow the pump(s) to run full cycles without damage; otherwise, they should be turned Off at their circuit breaker or disconnect switch.

When the pumps are turned Off, the starters for each pump and their run indicator pilot lights can be observed. If each pump control circuit is isolated by turning Off its circuit breaker or disconnect, the pump leads may have to be disconnected, and the circuit breaker or disconnect left On to observe starter operation while testing.

Purged air bubbler control systems can be checked through simulation by someone who fully understands their operation. The bubbler line to the wet well must be disconnected and a needle type cut-off valve installed on the line from the controls. Some units may have factory-installed shut-off and bleed valves.

With the test valve open completely, the level gauge should read zero to simulate a low wet well level. As the test valve is slowly closed, creating a back pressure on the controls, the gauge will indicate the simulated level and the pumps should start as the required levels are reached. When the test valve is slowly opened, lowering the back pressure on the controls, the pumps should turn Off.

This procedure may be repeated as required to adjust operating levels and check alternation with each duty cycle. During testing, the actual wet well level should be checked to ensure the pumps do not pump so far down as to damage them.

Other Checks

During the check-out of control systems,

Installation

the pump run pilot light indicators, elapsed time meters, and any other associated controls should be reviewed for proper operation. The alarm levels, when activated, should operate all devices in the system. These may include an alarm light, pilot light, audible horn or bell, and a relay to provide remote signal contacts or telemetering contacts.

All the alarm devices should be checked for proper operation, including audible alarm silencing circuits or switches, and any reset pushbutton, if used.

If any special control or alarm features are included in the system, they should be checked by simulation if actual conditions cannot be achieved. These may include seal failure indicators, motor thermal sensors, or telemetered incoming or outgoing signals.

Another device often used and not always checked is the phase monitor. Most have three functions — loss or low voltage on a single phase, loss or low voltage on all three phases, and phase rotation reversal.

By turning off the service voltage to the panel, disconnecting one of the sensing leads to the phase monitor, and then insulating it when the power is turned back onto the panel, the pumps should not operate in either the Hand or Automatic mode. Reversing the above procedure, the lead should then be reconnected. This will check the loss of a single-phase feature.

On some units, if the service nominal voltage is low enough, the set point on the monitor can be raised sufficiently above nominal voltage to cause the running pump or pumps to drop out. This will check the three-phase low voltage setting of the phase monitor.

By turning off the service voltage and reversing two of the sensing leads on the phase monitor (to induce improper rotation), and then restoring the service voltage, the pumps should not operate in either the Hand or Automatic mode. Reversing the above procedure, the leads should then be reconnected. This will check the phase reversal feature of the phase monitor.

These checks are important, since they offer the protection needed during normal operation of the system. The last is important if the service is ever disconnected and reconnected, to prevent the pumps from being run in reverse rotation. If a station is provided with a power plug for the use of a portable standby generator, this reverse phase feature is essential to ensure that the generator phase rotation is matched with the normal service phase rotation.

Final Tests

After the above tests have been completed, the panel should be thoroughly checked to ensure that all wires have been reconnected properly, all switches or jumpers used for simulations have been removed, and all circuits restored to normal operation.

With the service voltage turned Off, all wire terminations should be rechecked for tightness and the panel cleaned to ensure a good, maintainable environment. The panel should now be ready for continuous automatic operation.

The run time on elapsed time meters should be read and recorded for each pump. These readings should be the start of a permanent maintenance file. Periodic readings should be taken and recorded as part of a good preventive maintenance program.

With these readings, the alternating duty cycles can be checked to ensure operation of the alternator. Many pump and motor maintenance schedules are based on hours of operation. Actual automatic cycles are monitored to be sure they are not excessive. For submersible motors, a maximum of 10 to 15 starts per hour are acceptable in terms of equipment life.

Over a period of time, this record may show system demand trends and indicate the need for larger pumps. By calculating the pumping rate of each pump and the hours of run time, the gallons of water pumped for a given time period can be estimated. This may indicate problems with increased water infiltration into the system due to cracked or broken lines, based on an increase in gallons pumped with no other

Installation

substantial changes within the system.

Make periodic checks during the first few days of operation of a new system. This can uncover unforeseen problems and help the operator become familiar with duty cycles and other characteristics of the pumping station. This special attention can help avoid future problems.

Chapter 7
Operation and Maintenance

Regular inspection and preventive maintenance ensure continued, reliable operation of the entire submersible pumping system. All stations, pumps and operating equipment should be inspected at least once a year, and more frequently under severe operating conditions. One of the major advantages of a submersible station is the ability of the service technician to handle most maintenance and service on-site, without entering the wet well. All equipment in the station should be backed by manufacturers' service manuals. This material should be carefully read, filed, and should be consulted whenever servicing is required.

Safety Precautions

To minimize the risk of accidents in connection with service work, the following rules — as well as all applicable laws, regulations and manufacturers' recommendations — must be followed. *Note and read all safety precautions before performing any operation or maintenance procedure.*

- Be aware of health hazards. Observe strict cleanliness.
- Be aware of the risk of electrical accidents.
- Check the explosion risk before welding or using electric hand tools in or near the station. Never weld or use electrical tools in the wet well after it has been in operation.
- Make sure that all lifting equipment, when used, is in good condition.
- Provide a suitable barrier around the work area — for example, a guard rail.
- Make sure that all personnel have a clear path of retreat.
- Use safety helmets, safety goggles and protective shoes or boots.
- All personnel working with sewage systems must be vaccinated against any diseases that can occur.
- Never work alone. If there is a reason to enter the wet well, use a lifting harness and safety line.
- Before entering the wet well, make sure there is sufficient oxygen and that there are no poisonous gases present.

Since sewage pumps are designed for use in liquids which can be hazardous to the health, make sure that all equipment has been thoroughly cleaned.

To prevent injury to the eyes and skin, observe the following rules:

1. Always wear goggles and rubber gloves.
2. Wash and rinse the pump thoroughly with clean water before starting work.
3. Wash and rinse any components in water after disassembly and then dry thoroughly.

If you get hazardous chemicals in your eyes, rinse them immediately with running water for 15 minutes, and hold your eyelids apart with your fingers. Contact a doctor *immediately.*

If you get hazardous chemicals on your skin, remove contaminated clothes, wash your skin with soap and water, and seek medical attention *immediately.*

Recommended Inspections

Before starting work on any pump, make

Operation

sure it is isolated from the power supply and cannot be energized. This applies to the control circuit as well.

One method is to tag and lock the control panel to let other personnel know that you are working on the station. Keep in mind that some systems have an override switch at the treatment plant or other buildings. Make sure that this switch is also Off and tagged at the other building before you start working on the station.

After the pump(s) have been isolated from the power supply and pulled to the top of the station, the following inspection guidelines should be followed. The appropriate manufacturer's service manuals should be consulted in all cases.

Visible Parts On Pump And In Station

1. Check for vandalism or other station damage.
2. Make certain the access cover works properly. Check the hold-open device to ensure that it is engaged.
3. Make certain that the guide rails are completely vertical.
4. Check condition of the lifting eye, chains, hooks, and wire ropes.
5. Make certain that all screws, bolts and nuts are tight.
6. Replace or repair worn or damaged parts.

Pump Casing and Impeller

1. If the clearance between the impeller skirt and the pump casing or wear ring exceeds the manufacturer's recommendations, it may be necessary to adjust the impeller or replace the wear rings.
2. Wear on the outlet flange from the pump casing usually causes corresponding wear on the discharge connection.
3. Follow the manufacturer's instructions for disassembly, inspection and reassembly of the impeller and volute. When it is disassembled, check the motor shaft, impeller and volute bore for wear or damage.
4. Follow the manufacturer's instructions for disassembly, inspection and reassembly of the shaft seal. It must be clean and properly seated before reassembly.
5. Always replace worn or damaged parts.

Electrical Insulation

Perform megger (insulation resistance) test between the pump motor leads and pump casing. A low — 20 megohms or less — reading indicates moisture entry into the motor chamber or power cord, or other deterioration of the insulation system. Such problems should be corrected before a major breakdown occurs.

Oil Quantity and Condition

A. Oil-Filled Motors

1. *Caution:* If there has been any leakage, the motor housing may be under pressure. Hold a rag over the inspection plug to prevent splatter when loosening the plug.
2. Check the oil in both the motor housing and the seal cavity through the oil inspection plugs. The oil level may be low or emulsified (cream-like), which indicates that water has entered the cavity and a leak is present. One possible cause is an inspection plug which is not sufficiently tight. Check the sealing surface of the motor housing, the cable entry, and the condition of the shaft seal. Whatever the problem, correct it and make certain that the oil is refilled to the proper level.

B. Non Oil-Filled Motors

1. If there is any liquid in the motor housing, a leak is present and all sealing faces should be checked as previously mentioned under Oil-Filled Motors.

Cable Entry

1. Make certain that the cable connection is tight.
2. If the cable entry leaks, it may be necessary to replace the cable seal. See manufacturer's manual for instructions.
3. When refitting a cable which has been used before, even when the jacket is un-

Operation

damaged, always cut off a short piece of the cable so that the cable entry seal does not close around it at the same point.

4. If the outer jacket of the cable is damaged, replace the cable. Make sure the cable has no sharp bends and is not pinched.

Controls

1. Check liquid level sensors throughout their entire range of operation. Clean, adjust, replace and repair damaged equipment. Follow the manufacturer's instructions.

2. The same procedure should be used for checking the balance of the control system. In particular, check signals and the tripping function, and make sure that the relays, lamps, fuses and connections are intact. Replace all inoperative equipment.

Piping and Valves

1. Repair all flaws, and replace inoperative equipment.

Fault Tracing

A voltmeter, ohmmeter, and ammeter — together with the job wiring diagram — are required to test, measure and carry out fault tracing on electrical equipment.

Fault tracing must always be performed with the power supply disconnected and locked Off, except for those checks which can be performed only with power.

Electrical work must be performed by qualified electricians and all local, state, and national safety regulations be followed. Observe the recommended safety precautions previously mentioned in this chapter.

Major Servicing

Submersible sewage pumps can be serviced in the field at qualified facilities. If the pump is still in warranty, it should be serviced by an authorized shop.

Manufacturer's service manuals provide detailed instructions for replacement of impellers, stators, seals and bearings.

To facilitate field maintenance and service, many manufacturers provide a list of authorized service facilities, recommended spare parts, and the maintenance equipment required.

Trouble Checklist

Below is a list of common problems and their probable causes. These are general guidelines only. Consult the specific manufacturer's manual for detailed instructions.

Pump Operations

Problem	Possible Cause	Remedy
1. Pump will not start.	No power to motor.	Check for blown fuse or open circuit breaker.
	Selector switch may be in the Off position.	Turn to On position.
	Control circuit breaker may be tripped.	Reset the circuit breaker.
	Overload heater in starter may be tripped.	Push to reset.
	Overload heater in starter may be burnt out.	Replace the heater.

Operation

Problem	Possible Cause	Remedy
2. Pump will not start and overload heaters trip.	May be improperly grounded.	Turn power Off and check motor leads with megger or ohmmeter.
	Motor windings may be imbalanced.	Check resistance of motor windings. If three-phase, all phases should show the same reading.
	Impeller may be clogged, blocked or damaged.	If no grounds exist and the motor windings check out satisfactorily, remove pump from the well and check for impeller blockage.
3. Pump runs but will not shut off.	Pump may be air-locked.	Turn pump Off for several minutes, then restart.
	Lower level switch may be locked in closed position.	Check to be certain the level control is free.
	Selector switch may be in the Hand position.	Switch to Auto position.
4. Pump does not deliver proper capacity.	Discharge gate valve may be partially clogged.	Open and unclog valve.
	Check valve may be partially clogged.	Valve must be cleared. If there is an outside lever, move it up and down.
	Discharge line may be clogged.	Use a sewer cleaner or high-pressure hose to clear obstruction.
	Pump may be running in the wrong direction.	Low speed pumps can operate in reverse direction with little noise or vibration. See Chapter 6 for methods of establishing and correcting rotation.

Operation

Problem	Possible Cause	Remedy
	Discharge head may be too high.	Check total head with gauge when pump is operating. Compare against original design and previous operating records. If pump has been in service for some time and capacity falls off, remove pump and check for wear or clogged impeller.
5. Motor stops and then restarts after short period but overload heaters in starter do not trip.	Heat sensors in the motor may trip due to excessive heat.	Impeller may be partially clogged — resulting in the sustained overload, though not high enough to trip the overload heater switch.
	Motor may be operating out of liquid due to a failed level control.	Check locations and operations of level controls.
	Pump may be operating on a short cycle.	The wet well may be too small or water may be repeatedly returning to the well due to a leaking check valve. Both must be checked.

Control Panels

Problem	Possible Cause	Remedy
1. Pumps 1 and 2 will not run in Hand or Auto positions. Run lights not On.	Service voltage not On to panel.	Turn On and check for proper voltage.
	Main or control circuit breakers tripped or turned Off. Main or control circuit fuses blown.	Turn On or reset and turn On all circuit breakers. Check and replace any blown fuse.
	Motor heat sensor connections not made properly.	Check motor heat sensor connections and correct.

Operation

Problem	Possible Cause	Remedy
2. Pumps 1 and 2 will not run in Hand position. Run lights are On.	Motors not wired properly.	Check and correct connections and cables to panel.
	Incorrect voltage starter coils.	Check and correct starter coils voltage rating to match control circuit voltage.
3. Pump 1 or 2 will not run in Hand position. Run light not On. One pump operates in Hand position.	Pump circuit breaker tripped or turned Off.	Turn On or reset and turn On breaker.
	Pump circuit fuse blown.	Check and replace any blown fuses.
	Motor starter overload tripped.	Reset overload after checking motor.
	Motor heat sensor circuit open or not properly connected.	Check continuity of motor heat sensor. Correct connections.
4. Pumps 1 and 2 will not run in Auto position.	Level in wet well not high enough to turn on pumps.	Fill or allow wet well to fill to required levels.
	Level float switches may be incorrectly connected or failed.	Check and connect each float correctly or replace if required.
	Air bubbler supply may be off or failed.	Check and ensure air supply is On, bubbler line is working in wet well and has no leaks.
	Pressure switches or sensors may not be adjusted or sequenced properly.	Check and adjust pressure switches to correct levels and sequence.
	Relay or other control device failed.	Check and replace any control relay, alternator or other device with defective coil or contacts.

Operation

Problem	Possible Cause	Remedy
5. Alarm light and/or audible alarm turns On with both pumps running.	No probable panel problem.	None.
	Possible system problem — i.e., discharge line clogged.	Check and clear check valve or line of obstruction.
	Temporary high level condition after power failure or influent surge.	Monitor station operation until "High Level" is reduced.
6. Alarm light and/or audible alarm turns On, with one or both pumps not running.	Test Hand operation of pump not running and refer to Problem #3.	See Remedy for #3.
	Refer to Problem #4-B, D or E.	See Remedy for #4-B, 4-D or 4-E.
7. Circuit breaker tripped for motor power.	Motor not wired properly.	Check and correct connection to panel.
	Short in pump cable, wiring or motor.	Disconnect motor and check wiring. Check motor for shorts or grounds.
	Size of breaker too small and/or ambient heat problem.	Check and correct breaker size for motor and/or provide ventilation or compensation for ambient heat.
8. Blown fuse for motor power.	Same as Problem #7-A, B or C.	See Remedy #7.
9. Pumps do not alternate.	Defective alternator relay.	Check and replace alternator.
	Improper sequencing of float switches or pressure sensors.	Check and correct sequence of controls to insure Off, lead, lag sequence.
10. Indicator pilot light not On at the time of a function or alarm condition.	Pilot light bulb failed.	Replace lamp with correct voltage replacement.

GLOSSARY OF SYSTEM TERMS

ACCESS COVER — A removable device to provide access to wet well and/or valve box. Includes frame, doors, and accessories.
BEST EFFICIENCY POINT (BEP) — The combination of head and flow at which a given pump operates most efficiently.
BRAKE HORSEPOWER (BHP) — The horsepower required by the pump; pump input.
CAPACITY — The quantity of liquid that can be contained, or the rate of liquid flow that can be carried.
CAVITATION — The action resulting from forcing a flowing stream to change direction, in which reduced internal pressure causes dissolved gases to expand, creating negative pressure. Cavitation frequently causes pitting of the hydraulic structure affected.
CLEAR WATER — Treated, filtered water; the discharge from a water treatment plant.
CLOSE-COUPLED — A pump directly connected to its power unit without any reduction gearing or shafting.
CYCLE TIME — The total time period from when the pump turns On to when it turns Off.
DISCHARGE — The flow or rate of flow from a pump or pumping system.
DISCHARGE PIPE — The pipe that exits the wet well or valve box.
DRY WELL — A dry compartment in a pumping station, near or below pumping level, where the pumps are located.
DUPLEX — A pumping station containing two pumps.
DYNAMIC HEAD — The head (or pressure) against which a pump works.
EDDIES — A circular movement occurring in flowing water, caused by currents set up on the water by obstructions.
EFFLUENT — Wastewater or other liquid, partially or completely treated, flowing out of a septic tank or treatment plant.
FLOOD-PRONE — An area subject to frequent flooding.
FORCE MAIN — A pressure pipe joining the pump discharge at a water or wastewater pumping station with a point of gravity flow.
FREE BOARD — The vertical distance between the normal maximum level of liquid in a septic tank and the top of the tank.
FRICTION LOSS — The head loss of liquid flowing in a piping system as the result of the disturbances set up by the contact between the moving liquid and the system components.
GRINDER PUMPS — Specialized submersible pumps which macerate sewage. Used in pressurized systems.
GROUND WATER — Subsurface water occupying the saturation zone. In a strict sense, the term applies only to water below the water table.
GUIDE RAIL SYSTEM — A device which allows the pump-motor unit to be installed in or removed from the wet well, without disconnecting any piping and without anyone having to enter the wet well.

Glossary

H₂₀ WHEEL LOADING — Refers to the type of construction used in the fabrication of access doors or covers for wet wells or valve boxes that must withstand vehicular traffic. Rating is 16,000 lbs. per sq. ft. live load.

HEAD — The height of the free surface of fluid above any point in a hydraulic system; a measure of the pressure or force exerted by the fluid.

HYDRAULIC GRADIENT — The slope of the hydraulic grade line, the rate of change of pressure head; the ratio of the loss in the sum of the pressure head and position head to the flow distance.

IMPELLER — A rotating set of vanes designed to impel rotation of a mass of fluid.

INFILTRATION — The quantity of ground or near-surface water that leaks into a pipe or wet well through joints, porous walls, or breaks.

IN FLOW — The extraneous flow which enters a sanitary sewer from sources other than infiltration.

INTAKE — The flow or rate of flow into a pump or pump station.

LIFT STATION — A structure that contains pumps and appurtenant piping, valves, and other mechanical and electrical equipment for pumping water or wastewater. Also called *pumping station*.

LIQUID LEVEL CONTROLS — In-station devices which start or stop the pump when contacted by liquid or losing contact with liquid within the well. Connected to control panel. See *Chapter 4*.

LAG PUMP — A succeeding or backup pump in a pump system. Control sytems usually alternate pump operation.

LEAD PUMP — The first pump to start in a pump cycle.

NET POSITIVE SUCTION HEAD (NPSH) — The net positive suction head is the total suction head in feet of liquid absolute determined at the suction nozzle and the referred datum less the vapor pressure of the liquid in feet absolute.

NET POSITIVE SUCTION HEAD AVAILABLE (NPSHA) — The absolute pressure of the liquid at the inlet of the pump.

NET POSITIVE SUCTION HEAD REQUIRED (NPSHR) — Based on the need of a specific pump. Remains unchanged for a given head, flow, rotational speed and impeller diameter; changes with wear and liquids.

NON-CLOG — A pump designed to pass solids of a specific size. For example, a submersible pump with a 4-in. discharge may be capable of passing 3-in. spherical solids.

PEAK FACTOR — A variable multiplier used with average flow to determine required pump capacity for wastewater lift stations or potable water booster stations. Variation is determined by the size and type of facility.

PEAK FLOW — Maximum flow.

PUMP CASING — See *Volute*.

PUMP DISCHARGE SIZE — The nominal inside diameter of the discharge opening of a pump. On small sewage pumps, this may have a relationship to solids capability.

PUMP RELEASE SYSTEM — See *Guide Rail Systems*.

RAW WATER — Untreated water; usually the water entering the first treatment unit of a water treatment plant.

SANITARY WASTEWATER — Wastewater discharging from the sanitary conveniences of dwellings, including apartment houses, hotels, office buildings, industrial plants, or institutions.

Glossary

SCUM — The extra or foreign matter which rises to the surface of a liquid and forms a layer or film.

SEALING FLANGE — The connection between the pump discharge and force main when used with guide rail systems.

SEDIMENTATION — The process of subsidence and decomposition of suspended matter carried by water, wastewater or other liquids, by gravity. It is usually accomplished by reducing the velocity of the liquid below the point at which it can transport the suspended material.

SEPTIC TANK — An underground vessel for treating wastewater for a single dwelling or building by a combination of settling and anaerobic digestion. Effluent is usually disposed of by leaching. Settled solids are pumped out periodically and hauled to a treatment facility for disposal.

SEWAGE — Household or commercial wastewater that contains human waste. Distinguished from industrial wastewater.

SIMPLEX — A pumping station containing one pump.

SLUDGE — The accumulated solids which separate from liquids, such as water or wastewater, during processing.

SIPHON — The potential for atmospheric pressure to force a liquid through an inverted "U"-shaped tube from one point to another lower point over the barrier created by the inverted "U".

SOLIDS-HANDLING — The capability of a pump to pass solids of a specific size, such as 3-in. spherical solids.

SPECIFIC SPEED — An index number used to classify rotational pumps into three basic categories related to their vane and hydraulic configurations. $N_S = \dfrac{RPM \sqrt{GPM}}{H^{3/4}}$ The pump types are radial flow, mixed flow and axial flow.

N_S = 500 to 4,500 as radial flow
N_S = 4,500 to 10,000 as mixed flow
N_S = 10,000 to 15,000 as axial flow

STATIC ELEVATION (STATIC HEAD) — The vertical distance between the level of the source of supply and the high point in the force main or the level of the surface.

STORM WATER — Surface water from rain, snow, or melting ice which runs out from the surface of a drainage area. It is normally collected in sewers separate from the sanitary sewers, and receives minimal, if any treatment, prior to discharge to a receiving water. When collected in a combined sewer system, the resulting mixture of sewage and storm water is called combined wastewater.

SUBMERGENCE — The number of feet or inches of fluid above the center line of the volute on a vertical submersible pump. Requirements will vary depending on the nature of the fluid and the hydraulic configuration of the sump.

SUBMERSIBLE PUMPS — Submersible wastewater pumps are vertical, close-coupled, extra-heavy duty pump-motor units which are designed to operate under the liquid they are pumping. They are non-clogging, usually have a 3-in. or larger discharge, and are also called *submersible sewage pumps*. See also *Grinder Pumps*.

SUMP — See *Wet Well*.

SYSTEM HEAD CURVE — A graph showing the relationship of static head and friction head at various flow rates through a given piping system.

TOPOGRAPHY STUDY — A study that relates to the shape of the land surface and to the characteristics of the underlying soil and rocks.

Glossary

TOTAL DYNAMIC HEAD (TDH) — The difference between the elevation corresponding to the pressure at the discharge flange of a pump and the elevation corresponding to the vacuum or pressure at the suction inlet of the pump, corrected to the same datum plane, plus the velocity head at the discharge flange on the pump, minus the velocity head at the suction inlet of the pump.

TRIPLEX — A pumping station containing three pumps.

VALVE BOX — A metallic or concrete box or vault adjacent to the wet well containing valving, to allow access to valving for service and maintenance without having to enter the wet well.

VAPOR PRESSURE — That portion of pressure above atmospheric pressure that is required to maintain a liquid state for a given fluid at a specified temperature.

VELOCITY — The speed at which a liquid is moving. Usually referred to in feet per second.

VOLUTE — The casing of a centrifugal pump made in the form of a spiral or volute as an aid to the partial conversion of the velocity energy into pressure head as the water leaves the impeller.

VORTEX (VORTICES) — A revolving mass of water in which the streamlines are concentric circles and in which the total head of each streamline is the same.

WASTEWATER — The spent or used water of a community or industry which contains dissolved and suspended matter.

WATER HAMMER — A series of shocks within a piping system when the flow of liquid is stopped suddenly, with a sound like hammer blows.

WET WELL — A tank or pit which receives drainage, stores it temporarily, and from which the discharge is pumped.

GLOSSARY OF ELECTRICAL TERMS

ALTERNATING CURRENT (AC) — A current which reverses in regularly recurring intervals of time and which has alternative positive and negative values, and occurring a specified number of times per second. The number is expressed in cycles per second or hertz (Hz).

ALARM LIGHT — A light which is used to attract attention when a problem occurs in the system.

ALTERNATOR — A relay device designed for alternating the run cycle or duplexing action of two or more motors automatically. There are two basic types. One mechanically changes its contacts each time the operating coil is de-energized. The second is a solid state unit with an output relay. The alternator is used in the automatic control circuit to the motor starters to rotate the duty cycle of each motor.

AMBIENT TEMPERATURE — Temperature of the surroundings in which the equipment is used or operated.

AMMETER — Meter for measuring the current in an electrical circuit, measured in amperes.

AMPERE — The unit of electric current flow. One ampere will flow when one volt is applied across a resistance of one ohm.

AUDIBLE ALARM — Horn, siren, bell or buzzer which is used to attract the attention of the operator when a problem occurs in the system.

AUXILIARY CONTACTS — Contacts of a switching device in addition to the main current contacts that operate with the movement of the latter. They can be normally open (NO) or normally closed (NC) and change state when operated.

CAPACITOR — A device which introduces capacitance into an electrical circuit. The capacitor, when connected in an alternating current circuit, causes the current to lead the voltage in time phase. The peak of the current wave is reached ahead of the peak of the voltage wave. This is the result of the successive storage and discharge of electric energy.

CIRCUIT BREAKER — A mechanical switching device capable of making, carrying and breaking currents under normal conditions. Also making, carrying for a specific time, and automatically breaking currents under specified abnormal circuit conditions, such as those of short circuit. Circuit breakers have an ampere trip rating for normal thermal overload protection and a maximum magnetic ampere interrupting capacity (AIC) for short circuit protection.

COMMERCIAL POWER — The term applied to power furnished by an electric power utility.

CONDENSATION HEATER — A device that warms the air within an enclosure and prevents condensation of moisture during shut-down periods. Also known as a *space heater*.

CONDUCTOR — A wire, cable or bus bar designed for the passage of electrical current.

CONTACTOR — An electro-mechanical device that is operated by an electric coil and allows automatic or remote operation to repeatedly establish or interrupt an electrical power circuit. A contactor provides no overload protection as required for motor loads.

Glossary

CONTACTS — Devices for making and breaking electrical circuits, which are a part of all electrical switching devices.

CURRENT — The amount of electricity measured in amps which is flowing in a circuit.

CYCLE — A given length of time — see *Alternating Current*. In the U.S., most electric current is 60 cycle.

CYCLE TIMER — A timer that repeatedly opens and closes contacts according to pre-set time cycles.

DELTA CONNECTION — A common three-phase connection shaped schematically like the Greek Delta. The end of one phase is connected to the beginning of the next phase, or vice versa.

DEVICE, ELECTRICAL — A unit of an electrical system that is intended to carry but not utilize electric energy.

DISCONNECTING MEANS (DISCONNECT) — A device or group of devices, or other means whereby all the ungrounded conductors of a circuit can be disconnected simultaneously from their source of supply.

ELAPSED TIME METERS — The recording of the time that each pump has run. One elapsed time meter is used per pump.

ELECTRIC UTILITIES — All enterprises engaged in the production and/or distribution of electricity for use by the public.

EMERGENCY POWER (ALTERNATE SOURCE OF POWER) — An independent reserve source of electric power which, upon failure or outage of the normal power source, provides electric power.

ENCLOSURE — The cabinet or special designed box in which electrical controls and apparatus are housed. It is required by the NEC to protect persons from live electrical parts and limit access to authorized personnel. It also provides mechanical and environmental protection. An enclosure should be designed to provide the required protection and sized to provide good, safe wire access and replacement of components. It can be manufactured of steel, galvanized or stainless steel, aluminum, or fiberglass.

EXPLOSION-PROOF MOTOR — A motor in a special enclosure. The purpose of the enclosure is twofold:
1) If an explosive vapor (gas) should explode inside the motor, the frame of the motor will not be affected.
2) The enclosure is so constructed that no such explosion will ignite vapors outside the motor.

FACTORY MUTUAL (FM) — Independent U.S. agency which tests for safety.

FREQUENCY — The number of complete cycles of an alternating voltage or current per unit of time, usually per second, and expressed in cycles per second or hertz (Hz).

FULL LOAD CURRENT — The greatest current that a motor or other device is designed to carry under specific conditions; any additional is an overload.

FUSE — An over-current protective device which consists of a conductor that melts and breaks when current exceeds rated value beyond a predetermined time.

GENERAL PURPOSE RELAY — A relay that is adaptable to a wide variety of applications as opposed to a relay designed for a specific purpose or specific application.

GENERATOR — A machine for converting mechanical energy into electrical energy or power.

GENERATOR RECEPTACLE — A contact device installed for the connection of a plug and flexible cord to supply emergency power from a portable generator or other alternate source of power. Receptacles are rated in voltage, amps, number of wires, and by enclosure.

Glossary

GROUND — A connection, either intentional or accidental, between an electric circuit and the earth or some conducting body serving in place of the earth.

GROUND FAULT INTERRUPTION (GFI) — A unit or combination of units which provides protection against ground fault currents below the trip levels of the breakers of a circuit. The system must be carefully designed and installed to sense low magnitude insulation breakdowns and other faults that cause a fault ground current path. The GFI system must be capable of sensing the ground fault current and disconnecting the faulted circuit from the source voltage.

GROUNDED NEUTRAL — The common neutral conductor of an electrical system which is intentionally connected to ground to provide a current carrying path for the line to neutral load devices.

GROUNDING CONDUCTOR — The conductor that is used to establish a ground and that connects equipment, a device, a wiring system, or another conductor (usually the neutral conductor) with the grounding electrode.

HAND-OFF-AUTOMATIC (H.O.A.) — Selector switch determining mode of system operation. *H* is the hand mode only. *O* is system Off. *A* is automatic operation, normally with pump alternation.

HAZARDOUS LOCATIONS — Those areas where a potential for explosion and fire exist because of flammable gasses, vapors or finely pulverized dusts in the atmosphere, or because of the presence of easily ignitable fibers or flyings.

HERTZ (Hz) — A unit of frequency, also known as cycles per second.

HIGH POTENTIAL TEST — A test which consists of the application of a voltage higher than the rated voltage between windings and frame, or between two or more windings, for the purpose of determining the adequacy of insulating materials and spacing against breakdown under normal conditions. It is not the test of the conductor insulation of any one winding.

HORSEPOWER — The rate at which work is done. It is the result of the work done (stated in foot-pounds) divided by the time involved.

IN-RUSH CURRENT — See *Locked Rotor Current*.

INTERLOCK — Interrelates with other controllers. An auxiliary contact. A device connected in such a way that the motion of one part is held back by another part.

INTRINSICALLY SAFE — A term used to define a level of safety associated with the electrical controls used in some lift stations. Intrinsically safe equipment and wiring is incapable of releasing sufficient electrical or thermal energy under normal or abnormal conditions to cause ignition of a hazardous atmospheric mixture — without the need for explosion-proof enclosures in the hazardous area. Any associated devices must be outside the hazardous area with an approved seal-off fitting used as an isolating barrier.

KILOWATT (KW) — A unit of measure of electrical power. One kilowatt equals 1000 watts. Used where larger units of electrical power are measured.

LIGHTNING ARRESTOR — A protective device for limiting surge voltages on equipment by discharging or bypassing surge current; it prevents continued flow of follow current to ground, and is capable of repeating these functions as specified. Also known as a *surge arrestor*.

LOCKED ROTOR CURRENT — The current drawn by an electrical motor at the instant of power application or start-up. Current diminishes as the motor starts, unless rotation is prevented due to a bound or "locked" rotating element.

LOCKOUT — A mechanical device which may be set to prevent the operation of a pushbutton or other device.

Glossary

MANUAL TRANSFER SWITCH — A switch designed so that it will disconnect the load from one power source and reconnect it to another source while at no time allowing both sources to be connected to the load simultaneously.

MEGGER OR MEGOHMETER — A high resistance range ohmeter utilizing a power source for measuring insulation resistance.

MEGOHM — A unit of resistance equal to one million ohms.

MOTOR CIRCUIT PROTECTOR — A molded case disconnect switch specifically designed for motor circuits. It has a trip unit that operates on the magnetic principle only, sensing current in each of the three poles with an adjustable trip point. It provides short circuit protection, required by the NEC. It differs from a standard breaker in that it does not have a thermal overload unit.

MOTOR EFFICIENCY — A measure of how effectively the motor turns electrical energy into mechanical energy. Motor efficiency is never 100%. It is a variable that depends on a given motor's performance. Tabulated at 100, 75 and 50% load. It is the ratio of power input to power output.

MOTOR, ELECTRIC — A rotating device which converts electrical power into mechanical power.

NEC — The National Electrical Code is the standard of the National Board of Fire Underwriters for electric wiring and apparatus, as recommended by the National Fire Protection Association.

NEMA — National Electrical Manufacturers Association, a non-profit trade association supported by the manufacturers of electrical apparatus and supplies. NEMA promulgates standards to facilitate understanding between the manufacturers and users of electrical products.

NFPA — National Fire Protection Association, of which the National Electric Code (NEC) is a chapter.

NEUTRAL — The point common to all phases of a polyphase circuit, a conductor to that point, or the return conductor in a single phase circuit. The neutral in most systems is grounded at or near the point of service entrance only and becomes the grounded neutral.

NORMALLY OPEN and NORMALLY CLOSED — The terms "Normally Open" and "Normally Closed" when applied to a magnetically operated switching device — such as a contactor or relay, or to the contacts thereof — signify the position taken when the operating magnet is de-energized. These terms pertain to all switches.

OHM — Unit of electrical resistance. One volt will cause a current of one ampere to flow through a resistance of one ohm.

OHMMETER — A device for measuring electrical resistance expressed in ohms.

OVERLOAD PROTECTION — Overload protection is the effect of a device operative on excessive current, but not necessarily on short circuit, to cause and maintain the interruption of current flow to the device being governed. Re-set may be manual or automatic.

OVERLOAD RELAY — A relay that responds to electric load and operates at a pre-set value of overload. The unit senses the current in each line to the motor and is either bi-metallic, melting alloy or solid state actuated. It may be of the non-compensated or ambient-compensated type, and of a standard or fast-trip design.

Glossary

PHASE MONITOR — A device in the control circuit of motors which monitors the three phase voltage and protects against a phase loss (single phasing), under voltage (brown outs) and phase reversal (improper phase sequence). Most are adjustable to set the nominal voltage and some have a LED indicator to indicate acceptable voltage and phase conditions. The output contacts are used to control the motor starters and provide signaling for telemetering.

PILOT DEVICE — Directs operation of another device:
  Float switch — A pilot device responding to liquid levels.
  Limit switch — A pilot device operated by response of a mechanical operation.
  Pressure switch — A pilot device operated in response to pressure levels.
  Temperature switch — A pilot device operated in response to temperature values.
  All of the above switches cause a contact change of the switch at pre-set or adjustable points.

PILOT LIGHT — A lamp available with various colored lens designed to operate on a control voltage. They are each turned On and Off to provide the required indication for specific functions or alarm conditions. They are available in various sizes and voltage ratings. They are each designed for a specific bulb style and base configuration and some have an integral transformer to allow the use of low voltage bulbs. Full voltage incandescent bulbs are most common, but neon bulbs are also used.

POWER FACTOR — The ratio of the true power to the volt-amperes in an alternating current circuit. Power factor is expressed in a percent of unity either lagging for inductive loads or leading for capacitive loads. Resistive loads produce a unity power factor.

PUSHBUTTON — Part of an electrical device, consisting of a button that must be pressed to effect an operation.

RATED VOLTAGE — The voltage of electrical apparatus at which it is designed to operate.

REDUCED VOLTAGE AUTO-TRANSFORMER STARTER — A starter that includes an auto-transformer to furnish reduced voltage for starting an alternating current motor. It includes the necessary switching mechanism. This is the most widely used reduced voltage starter because of its efficiency and flexibility.

RELAY — An electric device that is designed to interpret input conditions in a prescribed manner and, after specified conditions are met, to respond and cause contact operation or similar abrupt changes in associated electric control circuits.

RELAY, ELECTROMAGNETIC — A relay controlled by electromagnetic means, to open and close electric contacts.

RELAY, SOLID STATE — A completely electronic switching device with no moving parts or contacts.

REMOTE CONTROL — Control function initiation or change of electrical device from a remote point.

RESISTANCE — The non-reactive opposition which a device or material offers to the flow of direct or alternating current.

SAFETY SWITCH — An enclosed, manually-operated disconnecting switch, which is horsepower and current rated. Disconnects all power lines simultaneously.

SEAL FAILURE ALARM — The sensing and indication of the intrusion of water in the oil-filled seal chamber between the inner and outer shaft seal of a submersible pump.

SELECTOR SWITCH — A multi-position switch which can be set to the selected mode of operation.

SERVICE FACTOR — A safety factor designed and built into some motors which allows the motor, when necessary, to delivery greater than its rated horsepower.

Glossary

SINGLE PHASE — A circuit that differs in phase by 180 degrees. Single phase circuits have two conductors, one of which may be a neutral, or three conductors, one of which is neutral.

STANDBY POWER SUPPLY — The power supply that is available to furnish electric power when the normal power supply is not available.

STAR CONNECTION — Same as a "Y" or "Wye" connection. This three-phase connection is so called because, schematically, the joint of the "Y" points looks like a star.

STARTER — A device used to control the electrical power to motors and provide overload protection as required by the NEC. The starter can be operated manually, electrically, or by automatic pilot devices. A starter has two basic parts — a contactor for power switching and an overload relay for protection.

STARTING RELAY — A relay — actuated by current, voltage or the combined effect of current and voltage — which is used to perform a circuit-changing function in the primary winding of single phase induction motor within a pre-determined range of speed as the motor accelerates; and to perform the reverse circuit-changing operation when the motor is disconnected from the supply line. One of the circuit changes that is usually performed is to open or disconnect the auxiliary winding (starting) circuit.

SUBMERSIBLE MOTOR — A motor whose housing and terminal box is so designed that the motor can run underwater — completely submerged at an allowable temperature.

SWITCH — A device for making, breaking, or changing connections in a circuit.

TELEMETERING — The transmitting of alarm and control signals to and from remote lift station controls and a central monitoring location.

TERMINAL BLOCK — An insulating base equipped with terminals for connecting wires.

THREE PHASE CIRCUIT — A combination of circuits energized by alternating electromotive sources which differ in phase by one third of a cycle — that is, 120 degrees. A three-phase circuit may be three wire or four wire with the fourth wire being connected to the neutral point of the circuit which may be grounded.

TIME CLOCK — A device used to schedule electrical On and Off operations based on the time of day. A time clock is usually driven by a synchronous motor and must be manually set. Some clocks use 15 minute increments and some have up to six per day minimum 20 minute cycles.

TIME DELAY RELAY (TDR) — A device with either relay or solid state output contacts that performs a timing function upon energization or control signal.

TRANSDUCER — A device to condition and transform an analog signal to a specific variable output electrical signal proportional to the input signal. Typical inputs include variable pressure, level, voltage or current. Some common outputs are 0-1ma, 4-20ma, and various MVDC signals. A transducer must be specifically designed to be compatible with the input/output requirements of the total system.

TRANSFORMER — A static electric device consisting of a single winding, or two or more coupled windings, used to transfer power by electromagnetic induction between circuits at the same frequency, usually with changed values of voltage and current.

UL — Underwriters Laboratories, Inc. An independent, non-profit U.S. organization that tests products for safety.

VOLTAGE — The potential or electrical magnetic force (EMF) in an electrical circuit, similar to pressure in a water system.

VOLTMETER — An instrument for measuring voltage.

WATT — A unit of measure of electrical power.

WYE CONNECTION — See *Star Connection*.

APPENDIXES

ELECTRICAL DIAGRAM TERMINOLOGY

WIRING DIAGRAM

A WIRING DIAGRAM shows, as closely as possible, the actual location of all of the component parts of the device. The open terminals (marked by an open circle) and arrows represent connections made by the user.

Since wiring connections and terminal markings are shown, this type of diagram is helpful when wiring the device, or tracing wires when troubleshooting. Note that bold lines denote the power circuit, and thin lines are used to show the control circuit. Conventionally, in ac magnetic equipment, black wires are used in power circuits and red wiring is used for control circuits.

A wiring diagram, however, is limited in its ability to convey a clear picture of the sequence of operation of a controller. Where an illustration of the circuit in its simplest form is desired, the elementary diagram is used.

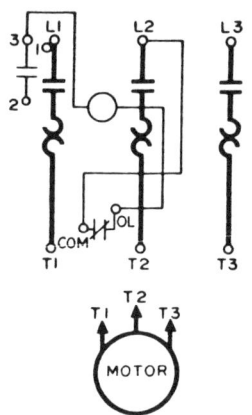

ELEMENTARY DIAGRAM

The elementary diagram gives a fast, easily understood picture of the circuit. The devices and components are not shown in their actual positions. All the control circuit components are shown as directly as possible, between a pair of vertical lines, representing the control power supply. The arrangement of the components is designed to show the sequence of operation of the devices, and helps in understanding how the circuit operates. The effect of operating various interlocks, control devices etc. can be readily seen — this helps in trouble shooting, particularly with the more complex controllers. This form of electrical diagram is sometimes referred to as a "schematic" or "line" diagram.

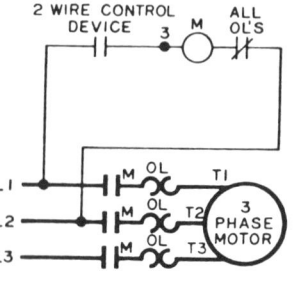

ELECTRICAL SYMBOLS

DISCONNECT	CIRCUIT INTERRUPTER	CIRCUIT BREAKER Thermal	LIMIT SWITCH				
			SPRING RETURN			MAINTAINED	
			Normally Open	Normally Closed	Neutral Position NP		
			Held Closed	Held Open			

LIQUID LEVEL		VACUUM & PRESSURE		TEMPERATURE ACTIVATED		FLOW (AIR, WATER, ETC.)	
Normally Open	Normally Closed	Normally Open	Normally Closed	Normally Open	Normally Closed	Normally Open	Normally Closed

PUSH BUTTONS					FOOT SWITCH	
Normally Open	Normally Closed	Double Circuit	Mushroom Head	Maintained	Normally Open	Normally Closed

H.O.A. SWITCH		LAMPS PUSH TO TEST DENOTE COLOR BY LETTER	TIME DELAY CONTACT			
HAND OFF AUTO	Hand-off-auto toggle selector switch		Normally Open	Normally Open OR	Normally Closed OR	Normally Closed OR

GENERAL CONTACTS		CONDUCTORS		MAGNET COIL	CONTROL TRANSFORMER	METER
Normally Open	Normally Closed	Not Connected	Connected		H1 H3 H2 H4 X2 X1	VM AM

GROUND	FULL WAVE RECTIFIER	HORN, SIREN	BELL, BUZZER	MOTOR 3 Phase	OVERLOAD RELAY Thermal	FUSE

AUTO TRANSFORMER	RESISTOR		LOCATION OF RELAY CONTACTS
	Adjustable RES	Fixed RES	ICR (2-3-4) 1 ICR 2 ICR 3 ICR 4 ICR — Numbers In Parenthesis Designate The Location Of Relay Contacts. A Line Underneath A Location Number Signifies A Normally Closed Contact

ELECTRICAL FORMULAE

FACTOR	NEW SYMBOLS ALTERNATING CURRENT	OLD SYMBOLS ALTERNATING CURRENT
Watts	A x V x PF (1-phase) A x V x 1.73 x PF (3-phase)	I x E x PF (1-phase) I x E x 1.73 x PF (3-phase)
Kilowatts	$\dfrac{A \times V \times PF}{1000}$ (1-phase) $\dfrac{A \times V \times 1.73 \times PF}{1000}$ (3-phase)	$\dfrac{I \times E \times PF}{1000}$ (1-phase) $\dfrac{I \times E \times 1.73 \times PF}{1000}$ (3-phase)
Amperes (when kW is known)	$\dfrac{kW \times 1000}{V \times PF}$ (1-phase) $\dfrac{kW \times 1000}{V \times 1.73 \times PF}$ (3-phase)	$\dfrac{KW \times 1000}{E \times PF}$ (1-phase) $\dfrac{KW \times 1000}{E \times 1.73 \times PF}$ (3-phase)
Kilovolt-amperes (kVA)	$\dfrac{A \times V}{1000}$ (1-phase) $\dfrac{A \times V \times 1.73}{1000}$ (3-phase)	$\dfrac{I \times E}{1000}$ (1-phase) $\dfrac{I \times E \times 1.73}{1000}$ (3-phase)
Frequency (hertz) Hz	$\dfrac{p \times RRM}{120}$	$\dfrac{p \times RPM}{120}$
Revolutions per minute	$\dfrac{Hz \times 120}{P}$	$\dfrac{f \times 120}{P}$
Power factor	$\dfrac{\text{Actual watts}}{A \times V} = \dfrac{kW}{kVA}$	$\dfrac{\text{Actual watts}}{I \times E} = \dfrac{KW}{KVA}$
Amperes when kilowatts is known	$\dfrac{kW \times 1000}{V \times PF}$ (1-phase)	$\dfrac{KW \times 1000}{E \times PF}$ (1-phase)
Amperes when kilovolt- amperes is known	$\dfrac{kVA \times 1000}{V}$ (1-phase)	$\dfrac{KVA \times 1000}{E}$ (1-phase)
Amperes when horsepower is known	$\dfrac{kW}{V \times PF \times \%Eff}$	$\dfrac{HP \times 746 \times \%Eff}{E \times PF}$

ELECTRICAL CONVERSION FORMULAE

DESIRED DATA	SINGLE PHASE	THREE PHASE
KILO VOLT-Amperes (KVA)	$\dfrac{\text{Volts} \times \text{AMPS}}{1000}$ or $\dfrac{\text{KW}}{\text{P.F.}}$	$\dfrac{1.73 \times \text{Volts} \times \text{AMPS}}{1000}$ or $\dfrac{\text{KW}}{\text{P.F.}}$
Kilowatts (KW)	$\dfrac{\text{Volts} \times \text{AMPS} \times \text{P.F.}}{1000}$ or $\text{KVA} \times \text{P.F.}$	$\dfrac{1.73 \times \text{Volts} \times \text{AMPS} \times \text{P.F.}}{1000}$ or $\text{KVA} \times \text{P.F.}$
Power Factor (P.F.)	$\dfrac{\text{KW}}{\text{KVA}}$	$\dfrac{\text{KW}}{\text{KVA}}$
Amperes — When KW is known	$\dfrac{\text{KW} \times 1000}{\text{Volts} \times \text{P.F.}}$	$\dfrac{\text{KW} \times 1000}{1.73 \times \text{Volts} \times \text{P.F.}}$
Amperes — When KVA is known	$\dfrac{\text{KVA} \times 1000}{\text{Volts}}$	$\dfrac{\text{KVA} \times 1000}{1.73 \times \text{Volts}}$
Frequency (Hertz)	$\dfrac{\text{Number of Poles} \times \text{R.P.M.}}{120}$	
Revolutions Per Minute (R.P.M.)	$\dfrac{\text{Hertz} \times 120}{\text{Number of Poles}}$	

INSULATION RESISTANCE READINGS

CONDITION OF MOTOR AND LEADS	OHM VALUE	MEGOHM VALUE
A new motor (without drop cable).	20,000,000 (or more)	20.0
A used motor which can be reinstalled in the well.	10,000,000 (or more)	10.0
MOTOR IN WELL. Ohm readings are for drop cable plus motor.		
A new motor in the well.	2,000,000 (or more)	2.0
A motor in the well in reasonably good condition.	500,000-2,000,000	0.5 -2.0
A motor which may have been damaged by lightning or with damaged leads. Do not pull the pump for this reason.	20,000- 500,000	0.02-0.5
A motor which definitely has been damaged or with damaged cable. The pump should be pulled and repairs made to the cable or the motor replaced. The motor will not fail for this reason alone, but it will probably not operate for long.	10,000- 20,000	0.01-0.02
A motor which has failed or with completely destroyed cable insulation. The pump must be pulled and the cable repaired or the motor replaced.	less than 10,000	0-0.01

Courtesy: Franklin Electric Company, Inc.

HAZARDOUS LOCATION BASICS

To do the job safely—you need the facts on the various types of Hazardous Locations.

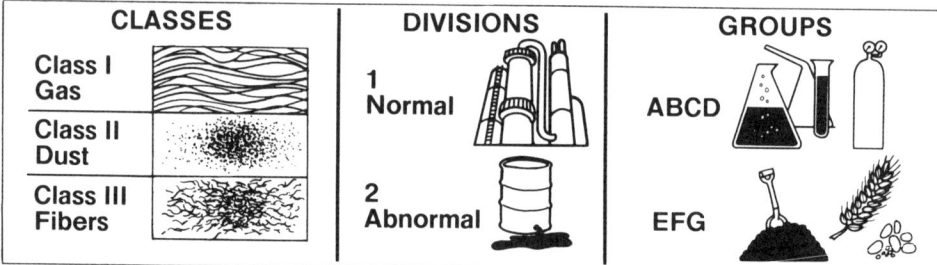

CLASSES	DIVISIONS	GROUPS
Class I Gas	1 Normal	ABCD
Class II Dust	2 Abnormal	EFG
Class III Fibers		

Definitions*

HAZARDOUS LOCATION—An area where the possibility of explosion and fire is created by the presence of flammable gases, vapors, dusts, fibers or flyings.

CLASS I (NEC-500-4)—Those areas in which flammable gases or vapors may be present in the air in sufficient quantities to be explosive or ignitable.

CLASS II (NEC-500-5)—Those areas made hazardous by the presence of combustible **dust**.

CLASS III (NEC-500-6)—Those areas in which there are easily ignitable fibers or flyings present, due to type of material being handled, stored, or processed.

DIVISION 1 (NEC-500-4, 5, 6)—Division One in the normal situation, the hazard would be **expected** to be present in everyday production operations or during frequent repair and maintenance activity.

DIVISION 2 (NEC-500-4, 5, 6)—Division Two in the abnormal situation, material is expected to be confined within closed containers or closed systems and will be present only through accidental rupture, breakage or unusual faulty operation.

GROUPS (NEC-500-2)—The gases and vapors of Class I locations are broken into four groups by the code. A, B, C and D. These materials are grouped according to . . . the ignition temperature of the substance, its explosion pressure and other flammable characteristics.

Class II—dust locations—groups E, F① and G. These groups are classified according to the **ignition temperature** and the **conductivity** of the hazardous substance.

SEALS (NEC-501-5 & 502-5)—Special fittings that are required either to prevent the passage of hot gases in the case of an explosion in a Class I area or the passage of combustible dust, fibers, or flyings in a Class II or III area.

ARTICLES 500 Through 503 (1984 NEC)—Explains in detail the requirements for the installation of wiring or electrical equipment in hazardous locations. These articles along with other applicable regulations; local governing inspection authorities, insurance representatives, and qualified engineering/technical assistance should be your guides to the installation of wiring or electrical equipment in any hazardous or potentially hazardous location.

Typical Class I locations:

- Petroleum refineries, and gasoline storage and dispensing areas.
- Industrial firms that use flammable liquids in dip tanks for parts cleaning or other operations.
- Petrochemical companies that manufacture chemicals from gas and oil.
- Dry cleaning plants where vapors from cleaning fluids can be present.
- Companies that have spraying areas where they coat products with paint or plastics.
- Aircraft hangars and fuel servicing areas.
- Utility gas plants, and operations involving storage and handling of liquified petroleum gas or natural gas.

Typical Class II locations:

- Grain elevators, flour and feed mills.
- Plants that manufacture, use or store magnesium or aluminum powders.
- Plants that have chemical or metallurgical processes... producers of plastics, medicines and fireworks, etc.
- Producers of starch or candies.
- Spice-grinding plants, sugar plants and cocoa plants.
- Coal preparation plants and other carbon-handling or processing areas.

Typical Class III locations:

- Textile mills, cotton gins, cotton seed mills and flax processing plants.
- Any plant that shapes, pulverizes or cuts wood and creates sawdust or flyings.

Note: fibers and flyings are not likely to be suspended in the air, but can collect around machinery or on lighting fixtures and where heat, a spark or hot metal can ignite them.

*These are simplified definitions — complete data available in the reference 1984 N.E.C. article.
①Class II Group F has been deleted from the 1984 N.E.C., the dusts previously listed in this group have been placed in group E or G, depending on their electrical resistivity.

KILLARK ELECTRIC MANUFACTURING COMPANY, P.O. Box 5325, St. Louis, MO 63101
KILLARK ELECTRIC LIMITED, 1051 Brevik Place, Mississauga, Ontario L4W 3R7, (416) 624-2401

LOADING: ACCESS COVERS

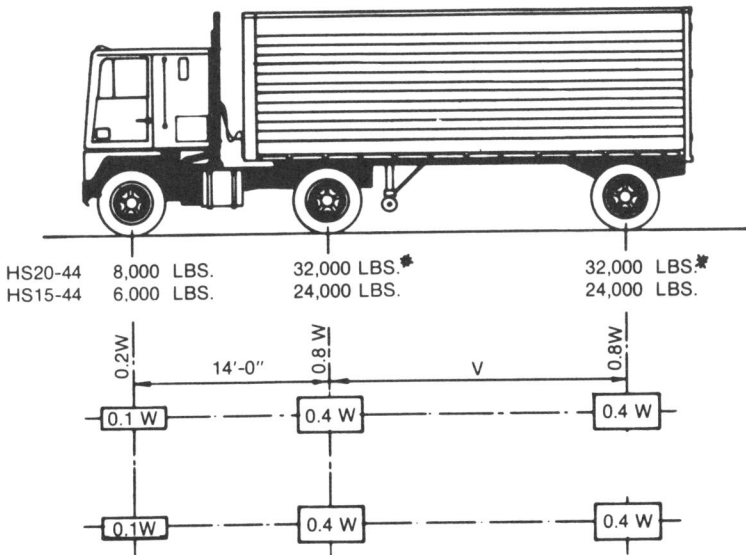

W = COMBINED WEIGHT ON THE FIRST TWO AXLES WHICH IS THE SAME AS FOR THE CORRESPONDING H (M) TRUCK.
V = VARIABLE SPACING — 14 FEET TO 30 FEET INCLUSIVE. SPACING TO BE USED IS THAT WHICH PRODUCES MAXIMUM STRESSES.

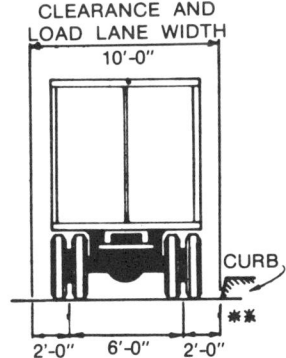

Figure 3.7.7A. Standard HS Trucks

*In the design of timber floors and orthotropic steel decks (excluding transverse beams) for HS 20 loading, one axle load of 24,000 pounds or two axle loads of 16,000 pounds each, spaced 4 feet apart may be used, whichever produces the greater stress, instead of the 32,000-pound axle shown.

**For slab design, the center line of wheels shall be assumed to be 1 foot from face of curb. (See Article 3.24.2).

Courtesy: AASHTO Standard Specification for Highway Bridges
— 13th Edition, 1983

METRIC EQUIVALENTS

WHEN YOU KNOW		YOU CAN FIND		IF YOU MULTIPLY BY
Length				
Inches	(in)	Millimeters	(mm)	25.4
Inches	(in)	Centimeters	(cm)	2.539 3
Feet	(ft)	Meters	(m)	0.304 8
Yards	(yd)	Meters	(m)	0.914 4
Miles (statute)	(mi)	Kilometers	(km)	1.609 344
Millimeters	(mm)	Inches	(in)	0.039 370 1
Meters	(m)	Feet	(ft)	3.280 840
Centimeters	(cm)	Inches	(in)	0.393 7
Meters	(m)	Yards	(yd)	1.093 61
Kilometers	(km)	Miles (statute)	(mi)	0.621 371 2
Area				
Square Inches	(in²)	Square Centimeters	(cm²)	6.451 6
Square Feet	(ft²)	Square Meters	(m²)	0.092 903 04
Square Yards	(yd²)	Square Meters	(m²)	0.836 127
Square Miles	(mi²)	Square Kilometers	(km²)	2.589 9
Acres		Hectares	(ha)	0.404 685 6
Square Centimeters	(cm²)	Square Inches	(in²)	0.155 000 3
Square Meters	(m²)	Square Feet	(ft²)	10.763 91
Square Meters	(m²)	Square Yards	(yd²)	1.195 99
Square Kilometers	(km²)	Square Miles	(mi²)	0.386 1
Hectares	(ha)	Acres		2.471 054
Liquid Volume				
Ounces	(oz)	Milliliters	(ml)	29.574
Pints	(pt)	Liters	(l)	0.473 2
Quarts	(qt)	Liters	(l)	0.946 3
Gallons	(gal)	Liters	(l)	3.785 412
Lb/Foot³	(lb/ft³)	Kilogram/Meter³	(kg/m³)	16.018 46
Lb/Gal	(lb/gal)	Kilogram/Meter³	(kg/m³)	119.826 4
Milliliters	(ml)	Ounces	(oz)	0.033 818
Liters	(l)	Pints	(pt)	2.113 271 3
Liters	(l)	Quarts	(qt)	1.056 747 3
Liters	(l)	Gallons	(gal)	0.264 172 0
Kilogram/Meter³	(kg/m³)	Lb/Foot³	(lb/ft³)	0.062 427 97
Kilogram/Meter³	(kg/m³)	Lb/Gal	(lb/gal)	0.008 345 4
Mass				
Ounces	(oz)	Grams	(g)	28.349 52
Pounds	(lb)	Kilograms	(kg)	0.453 592 4
Grams	(g)	Ounces	(oz)	0.035 275 97
Kilograms	(kg)	Pounds	(lb)	2.204 622

METRIC EQUIVALENTS (cont.)

WHEN YOU KNOW		YOU CAN FIND		IF YOU MULTIPLY BY
Force				
Foot Pound Force	(ft-lbf)	Joule	(J)	1.355 818
Newtons	(N)	Kilogram Force	(kgf)	0.101 971 6
Joule	(J)	Foot Pound Force	(ft-lbf)	0.737 562 1
Kilogram Force	(kgf)	Newtons	(N)	9.806 650
Power				
Horsepower	(hp)	Kilowatt	(kW)	0.745 699 9
Horsepower	(hp)	Watt	(W)	745.699 99
Kilowatt	(kW)	Horsepower	(hp)	1.341 022
Watt	(W)	Horsepower	(hp)	0.001 341 022
Pressure				
Lb/Inch²	(lb/in²)	Kilograms/meter²	(kg/m²)	703.069 7
Lb/Inch²	(lb/in²)	Kilopascal	(kPa)	6.894 7
Lb/Inch²	(lb/in²)	Meters of Water	(m)	0.704 089
Inches of Mercury	(inHg)	Kilograms/meter	(kg/m²)	345.3
Kilograms/meter²	(kg/m²)	Lb/inch²	(lb/in²)	0.001 422 3
Kilopascal	(kPa)	Lb/inch²	(lb/in²)	0.145 038 9
Meters of Water	(m)	Lb/inch²	(lb/in²)	1.420 274
Kilograms/meter²	(kg/m²)	Inches of Mercury	(inHg)	0.002 896
1 Atmosphere	14.7 lb/in²	Kilograms/Centimeter²	(kg/cm²)	1.033
1 Atmosphere	14.7 lb/in²	Kilograms/Meter²	(kg/m²)	10 335.0
1 Atmosphere	14.7 lb/in²	Kilopascal	(kPa)	101.35
1 Atmosphere	14.7 lb/in²	76.0 Centimeters Mercury	(cmHg)	5.17
1 Atmosphere	29.92 inHg	76.0 Centimeters Mercury	(cmHg)	2.54
Pascal	(Pa)	Newton/meter²	(N/m²)	1.0
Pascal	(Pa)	Kilogram Per Meter²	(kg/m²)	0.101 971 6
Temperature				
Degrees Fahrenheit	(°F)	Degrees Celsius	(°C)	(°F-32) 0.555 555
Degrees Celsius	(°C)	Degrees Fahrenheit	(°F)	(1.8 x °C) + 32
Velocity & Flow				
Feet Per Second	(fps)	Meter Per Second	(m/s)	0.304 8
Mile Per Hour	(mph)	Kilometer Per Hour	(km/h)	1.609 344
Gallons Per Minute	(gpm)	Liter Per Minute	(l/min)	3.785 412
Gallons Per Minute	(gpm)	Cubic Meter Per Minute	(m³/min)	0.003 785 412
Meter Per Second	(m/s)	Feet Per Second	(fps)	3.280 840
Kilometer Per Hour	(km/h)	Mile Per Hour	(mph)	0.621 371 2
Liter Per Minute	(l/m)	Gallons Per Minute	(gpm)	0.264 172
Cubic Meter Per Minute	(m³/min)	Gallons Per Minute	(gpm)	264.172 0

Electrical Terms: Amperes, Watts, Kilowatts, Volts and Ohms are the same in Metric.

FRICTION OF WATER IN PLASTIC PIPE
1 inch
C = 150

Flow US GPM	Schedule 40 1.049 in. I.D. Velocity ft/sec	Schedule 40 Head loss ft/100 ft	Schedule 80 0.957 in. I.D. Velocity ft/sec	Schedule 80 Head loss ft/100 ft	Schedule 120 0.915 in. I.D. Velocity ft/sec	Schedule 120 Head loss ft/100 ft	Flow US GPM
2	0.74	0.28	0.89	0.44	0.98	0.55	2
3	1.11	0.59	1.34	0.93	1.46	1.16	3
4	1.48	1.01	1.78	1.58	1.95	1.97	4
5	1.86	1.53	2.23	2.39	2.44	2.98	5
6	2.23	2.14	2.68	3.35	2.93	4.17	6
8	2.97	3.65	3.57	5.71	3.90	7.10	8
10	3.71	5.52	4.46	8.63	4.88	10.73	10
12	4.45	7.73	5.35	12.09	5.86	15.04	12
14	5.20	10.28	6.24	16.07	6.83	20.00	14
16	5.94	13.17	7.14	20.58	7.81	25.60	16
18	6.68	16.37	8.03	25.59	8.78	31.83	18
20	7.42	19.89	8.92	31.10	9.76	38.68	20
22	8.17	23.73	9.81	37.09	10.73	46.14	22
24	8.91	27.88	10.70	43.57	11.71	54.20	24
26	9.65	32.32	11.60	50.52	12.69	62.85	26
28	10.39	37.07	12.49	57.95	13.66	72.09	28
30	11.14	42.12	13.38	65.84	14.64	81.90	30
35	12.99	56.02	15.61	87.56	17.08	108.93	35
40	14.85	71.72	17.84	112.10	19.52	139.46	40
45	16.71	89.18	20.07	139.39	21.96	173.41	45

1.25 inch
C = 150

Flow US GPM	Schedule 40 1.380 in. I.D. Velocity ft/sec	Schedule 40 Head loss ft/100 ft	Schedule 80 1.278 in. I.D. Velocity ft/sec	Schedule 80 Head loss ft/100 ft	Schedule 120 1.230 in. I.D. Velocity ft/sec	Schedule 120 Head loss ft/100 ft	Flow US GPM
4	0.86	0.27	1.00	0.39	1.08	0.47	4
5	1.07	0.40	1.25	0.59	1.35	0.71	5
6	1.29	0.56	1.50	0.82	1.62	0.99	6
7	1.50	0.75	1.75	1.09	1.89	1.31	7
8	1.72	0.96	2.00	1.40	2.16	1.68	8
10	2.15	1.45	2.50	2.11	2.70	2.54	10
12	2.57	2.04	3.00	2.96	3.24	3.56	12
14	3.00	2.71	3.50	3.93	3.78	4.74	14
16	3.43	3.47	4.00	5.04	4.32	6.07	16
18	3.86	4.31	4.50	6.26	4.86	7.55	18
20	4.29	5.24	5.00	7.61	5.40	9.17	20
25	5.36	7.92	6.25	11.50	6.75	13.86	25
30	6.44	11.09	7.50	16.12	8.10	19.42	30
35	7.51	14.75	8.75	21.43	9.45	25.82	35
40	8.58	18.89	10.00	27.44	10.80	33.06	40
50	10.73	28.54	12.51	41.47	13.50	49.96	50
60	12.87	39.99	15.01	58.10	16.20	70.00	60
70	15.02	53.18	17.51	77.27	18.90	93.09	70
80	17.16	68.09	20.01	98.93	21.60	119.18	80
90	19.31	84.66	22.51	123.01	24.30	148.20	90

FRICTION OF WATER IN PLASTIC PIPE
1.5 inch
C = 150

Flow US GPM	Schedule 40 1.610 in. I.D.		Schedule 80 1.500 in. I.D.		Schedule 120 1.450 in. I.D.		Flow US GPM
	Velocity ft/sec	Head loss ft/100 ft	Velocity ft/sec	Head loss ft/100 ft	Velocity ft/sec	Head loss ft/100 ft	
4	0.63	0.13	0.73	0.18	0.78	0.21	4
5	0.79	0.19	0.91	0.27	0.97	0.32	5
6	0.95	0.27	1.09	0.38	1.17	0.44	6
7	1.10	0.35	1.27	0.50	1.36	0.59	7
8	1.26	0.45	1.45	0.64	1.55	0.76	8
9	1.42	0.56	1.63	0.80	1.75	0.94	9
10	1.58	0.69	1.82	0.97	1.94	1.14	10
12	1.89	0.96	2.18	1.36	2.33	1.60	12
14	2.21	1.28	2.54	1.81	2.72	2.13	14
16	2.52	1.64	2.90	2.31	3.11	2.73	16
18	2.84	2.04	3.27	2.87	3.50	3.39	18
20	3.15	2.47	3.63	3.49	3.89	4.12	20
22	3.47	2.95	3.99	4.17	4.27	4.91	22
24	3.78	3.47	4.36	4.89	4.66	5.77	24
26	4.10	4.02	4.72	5.67	5.05	6.69	26
28	4.41	4.61	5.08	6.51	5.44	7.67	28
30	4.73	5.24	5.45	7.39	5.83	8.72	30
32	5.04	5.90	5.81	8.33	6.22	9.82	32
34	5.36	6.60	6.17	9.32	6.61	10.99	34
36	5.67	7.34	6.54	10.36	6.99	12.22	36
38	5.99	8.11	6.90	11.45	7.38	13.50	38
40	6.30	8.92	7.26	12.59	7.77	14.85	40
42	6.62	9.76	7.63	13.78	8.16	16.25	42
44	6.93	10.64	7.99	15.02	8.55	17.71	44
46	7.25	11.55	8.35	16.30	8.94	19.23	46
48	7.56	12.50	8.71	17.64	9.33	20.80	48
50	7.88	13.48	9.08	19.02	9.71	22.43	50
55	8.67	16.08	9.99	22.69	10.69	26.76	55
60	9.46	18.89	10.89	26.65	11.66	31.43	60
65	10.24	21.90	11.80	30.91	12.63	36.45	65
70	11.03	25.12	12.71	35.45	13.60	41.80	70
75	11.82	28.54	13.62	40.27	14.57	47.49	75
80	12.61	32.16	14.52	45.38	15.54	53.52	80
85	13.40	35.98	15.43	50.77	16.51	59.87	85
90	14.18	39.99	16.34	56.43	17.49	66.55	90
95	14.97	44.20	17.25	62.36	18.46	73.55	95
100	15.76	48.60	18.16	68.57	19.43	80.87	100
110	17.34	57.97	19.97	81.79	21.37	96.46	110
120	18.91	68.09	21.79	96.08	23.32	113.31	120
130	20.49	78.96	23.60	111.42	25.26	131.40	130
140	22.06	90.56	25.42	127.79	27.20	150.70	140
150	23.64	102.89	27.23	145.18	29.14	171.22	150
160	25.22	115.94	29.05	163.60	31.09	192.93	160
170	26.79	129.70	30.86	183.01	33.03	215.83	170
180	28.37	144.17	32.68	203.42	34.97	239.90	180

FRICTION OF WATER IN PLASTIC PIPE
2 inch
C = 150

Flow US GPM	Schedule 40 2.067 in. I.D.		Schedule 80 1.939 in. I.D.		Schedule 120 1.875 in. I.D.		Flow US GPM
	Velocity ft/sec	Head loss ft/100 ft	Velocity ft/sec	Head loss ft/100 ft	Velocity ft/sec	Head loss ft/100 ft	
5	0.48	0.06	0.54	0.08	0.58	0.09	5
6	0.57	0.08	0.65	0.11	0.70	0.13	6
7	0.67	0.11	0.76	0.14	0.81	0.17	7
8	0.76	0.13	0.87	0.18	0.93	0.22	8
9	0.86	0.17	0.98	0.23	1.05	0.27	9
10	0.96	0.20	1.09	0.28	1.16	0.33	10
12	1.15	0.29	1.30	0.39	1.39	0.46	12
14	1.34	0.38	1.52	0.52	1.63	0.61	14
16	1.53	0.49	1.74	0.66	1.86	0.78	16
18	1.72	0.60	1.96	0.82	2.09	0.97	18
20	1.91	0.73	2.17	1.00	2.32	1.18	20
22	2.10	0.88	2.39	1.19	2.56	1.41	22
24	2.29	1.03	2.61	1.40	2.79	1.65	24
26	2.49	1.19	2.82	1.63	3.02	1.92	26
28	2.68	1.37	3.04	1.87	3.25	2.20	28
30	2.87	1.55	3.26	2.12	3.49	2.50	30
35	3.35	2.07	3.80	2.82	4.07	3.32	35
40	3.82	2.65	4.35	3.61	4.65	4.25	40
45	4.30	3.29	4.89	4.49	5.23	5.29	45
50	4.78	4.00	5.43	5.46	5.81	6.42	50
55	5.26	4.77	5.98	6.51	6.39	7.66	55
60	5.74	5.60	6.52	7.64	6.97	9.00	60
65	6.21	6.49	7.06	8.86	7.55	10.44	65
70	6.69	7.45	7.61	10.17	8.13	11.97	70
75	7.17	8.46	8.15	11.55	8.71	13.60	75
80	7.65	9.54	8.69	13.01	9.30	15.32	80
85	8.13	10.67	9.24	14.56	9.88	17.14	85
90	8.61	11.86	9.78	16.18	10.46	19.05	90
95	9.08	13.10	10.32	17.89	11.04	21.06	95
100	9.56	14.41	10.87	19.67	11.62	23.15	100
110	10.52	17.19	11.95	23.46	12.78	27.62	110
120	11.47	20.19	13.04	27.55	13.94	32.44	120
130	12.43	23.41	14.12	31.95	15.11	37.62	130
140	13.39	26.85	15.21	36.65	16.27	43.15	140
150	14.34	30.51	16.30	41.64	17.43	49.02	150
160	15.30	34.38	17.38	46.92	18.59	55.24	160
170	16.25	38.46	18.47	52.49	19.75	61.80	170
180	17.21	42.75	19.56	58.34	20.92	68.69	180
190	18.17	47.24	20.64	64.48	22.08	75.91	190
200	19.12	51.94	21.73	70.90	23.24	83.47	200
220	21.03	61.96	23.90	84.57	25.56	99.57	220
240	22.95	72.78	26.08	99.33	27.89	116.96	240
260	24.86	84.40	28.25	115.19	30.21	135.62	260
280	26.77	96.80	30.42	132.12	32.53	155.55	280
300	28.68	109.98	32.60	150.10	34.86	176.73	300

FRICTION OF WATER IN PLASTIC PIPE
2.5 inch
C = 150

Flow US GPM	Schedule 40 2.469 in. I.D.		Schedule 80 2.323 in. I.D.		Schedule 120 2.275 in. I.D.		Flow US GPM
	Velocity ft/sec	Head loss ft/100 ft	Velocity ft/sec	Head loss ft/100 ft	Velocity ft/sec	Head loss ft/100 ft	
8	0.54	0.06	0.61	0.08	0.63	0.08	8
10	0.67	0.09	0.76	0.12	0.79	0.13	10
12	0.80	0.12	0.91	0.16	0.95	0.18	12
14	0.94	0.16	1.06	0.21	1.10	0.24	14
16	1.07	0.20	1.21	0.28	1.26	0.30	16
18	1.21	0.25	1.36	0.34	1.42	0.38	18
20	1.34	0.31	1.51	0.42	1.58	0.46	20
22	1.47	0.37	1.67	0.50	1.74	0.55	22
24	1.61	0.43	1.82	0.58	1.89	0.64	24
26	1.74	0.50	1.97	0.68	2.05	0.75	26
28	1.88	0.58	2.12	0.77	2.21	0.86	28
30	2.01	0.65	2.27	0.88	2.37	0.97	30
35	2.35	0.87	2.65	1.17	2.76	1.30	35
40	2.68	1.11	3.03	1.50	3.16	1.66	40
45	3.02	1.39	3.41	1.86	3.55	2.06	45
50	3.35	1.68	3.78	2.26	3.95	2.51	50
55	3.69	2.01	4.16	2.70	4.34	2.99	55
60	4.02	2.36	4.54	3.17	4.74	3.51	60
65	4.36	2.74	4.92	3.68	5.13	4.07	65
70	4.69	3.14	5.30	4.22	5.52	4.67	70
75	5.03	3.56	5.68	4.79	5.92	5.31	75
80	5.36	4.02	6.06	5.40	6.31	5.98	80
85	5.70	4.49	6.43	6.04	6.71	6.69	85
90	6.03	4.99	6.81	6.72	7.10	7.44	90
95	6.37	5.52	7.19	7.43	7.50	8.22	95
100	6.70	6.07	7.57	8.16	7.89	9.04	100
110	7.37	7.24	8.33	9.74	8.68	10.78	110
120	8.04	8.50	9.08	11.44	9.47	12.66	120
130	8.71	9.86	9.84	13.26	10.26	14.68	130
140	9.38	11.31	10.60	15.21	11.05	16.84	140
150	10.05	12.85	11.35	17.29	11.84	19.13	150
160	10.72	14.48	12.11	19.48	12.63	21.56	160
170	11.39	16.20	12.87	21.79	13.42	24.12	170
180	12.06	18.00	13.63	24.22	14.21	26.81	180
190	12.73	19.90	14.38	26.77	15.00	29.63	190
200	13.40	21.88	15.14	29.43	15.79	32.58	200
220	14.74	26.10	16.65	35.11	17.36	38.86	220
240	16.08	30.65	18.17	41.24	18.94	45.65	240
260	17.42	35.55	19.68	47.82	20.52	52.93	260
280	18.76	40.77	21.20	54.85	22.10	60.71	280
300	20.10	46.32	22.71	62.31	23.68	68.98	300
350	23.45	61.61	26.49	82.88	27.62	91.74	350
400	26.80	78.87	30.28	106.10	31.57	117.45	400
450	30.16	98.07	34.06	131.93	35.52	146.04	450
500	33.51	119.18	37.85	160.33	39.46	177.47	500

FRICTION OF WATER IN PLASTIC PIPE
3 inch
C = 150

Flow US GPM	Schedule 40 3.068 in. I.D.		Schedule 80 2.900 in. I.D.		Schedule 120 2.800 in. I.D.		Flow US GPM
	Velocity ft/sec	Head loss ft/100 ft	Velocity ft/sec	Head loss ft/100 ft	Velocity ft/sec	Head loss ft/100 ft	
10	0.43	0.03	0.49	0.04	0.52	0.05	10
15	0.65	0.06	0.73	0.08	0.78	0.10	15
20	0.87	0.11	0.97	0.14	1.04	0.17	20
25	1.08	0.16	1.21	0.21	1.30	0.25	25
30	1.30	0.23	1.46	0.30	1.56	0.35	30
35	1.52	0.30	1.70	0.40	1.82	0.47	35
40	1.74	0.39	1.94	0.51	2.08	0.60	40
45	1.95	0.48	2.19	0.63	2.34	0.75	45
50	2.17	0.59	2.43	0.77	2.61	0.91	50
55	2.39	0.70	2.67	0.92	2.87	1.09	55
60	2.60	0.82	2.91	1.08	3.13	1.28	60
65	2.82	0.95	3.16	1.25	3.39	1.48	65
70	3.04	1.09	3.40	1.43	3.65	1.70	70
75	3.25	1.24	3.64	1.63	3.91	1.93	75
80	3.47	1.40	3.89	1.84	4.17	2.18	80
85	3.69	1.56	4.13	2.05	4.43	2.44	85
90	3.91	1.74	4.37	2.28	4.69	2.71	90
95	4.12	1.92	4.61	2.52	4.95	2.99	95
100	4.34	2.11	4.86	2.77	5.21	3.29	100
110	4.77	2.52	5.34	3.31	5.73	3.93	110
120	5.21	2.96	5.83	3.89	6.25	4.61	120
130	5.64	3.43	6.31	4.51	6.77	5.35	130
140	6.08	3.93	6.80	5.17	7.29	6.13	140
150	6.51	4.47	7.29	5.87	7.82	6.97	150
160	6.94	5.03	7.77	6.62	8.34	7.85	160
170	7.38	5.63	8.26	7.40	8.86	8.78	170
180	7.81	6.26	8.74	8.23	9.38	9.76	180
190	8.25	6.92	9.23	9.10	9.90	10.79	190
200	8.68	7.60	9.71	10.00	10.42	11.86	200
220	9.55	9.07	10.69	11.93	11.46	14.15	220
240	10.42	10.65	11.66	14.01	12.51	16.62	240
260	11.28	12.35	12.63	16.25	13.55	19.27	260
280	12.15	14.17	13.60	18.64	14.59	22.11	280
300	13.02	16.10	14.57	21.17	15.63	25.12	300
320	13.89	18.14	15.54	23.86	16.67	28.30	320
340	14.76	20.29	16.51	26.69	17.72	31.66	340
360	15.62	22.56	17.49	29.67	18.76	35.19	360
380	16.49	24.93	18.46	32.79	19.80	38.89	380
400	17.36	27.41	19.43	36.05	20.84	42.76	400
420	18.23	30.00	20.40	39.46	21.88	46.80	420
440	19.10	32.70	21.37	43.00	22.93	51.01	440
460	19.96	35.50	22.34	46.69	23.97	55.38	460
480	20.83	38.41	23.32	50.51	25.01	59.92	480
500	21.70	41.42	24.29	54.48	26.05	64.62	500
550	23.87	49.41	26.72	64.98	28.66	77.08	550

FRICTION OF WATER IN PLASTIC PIPE
4 inch
C = 150

Flow US GPM	Schedule 40 4.026 in. I.D.		Schedule 80 3.826 in. I.D.		Schedule 120 3.800 in. I.D.		Flow US GPM
	Velocity ft/sec	Head loss ft/100 ft	Velocity ft/sec	Head loss ft/100 ft	Velocity ft/sec	Head loss ft/100 ft	
20	0.50	0.03	0.56	0.04	0.57	0.04	20
30	0.76	0.06	0.84	0.08	0.85	0.08	30
40	1.01	0.10	1.12	0.13	1.13	0.14	40
50	1.26	0.16	1.40	0.20	1.41	0.21	50
60	1.51	0.22	1.67	0.28	1.70	0.29	60
70	1.76	0.29	1.95	0.37	1.98	0.38	70
80	2.02	0.37	2.23	0.48	2.26	0.49	80
90	2.27	0.46	2.51	0.59	2.55	0.61	90
100	2.52	0.56	2.79	0.72	2.83	0.74	100
110	2.77	0.67	3.07	0.86	3.11	0.89	110
120	3.02	0.79	3.35	1.01	3.39	1.04	120
130	3.28	0.91	3.63	1.17	3.68	1.21	130
140	3.53	1.05	3.91	1.34	3.96	1.39	140
150	3.78	1.19	4.19	1.53	4.24	1.58	150
160	4.03	1.34	4.47	1.72	4.53	1.78	160
170	4.28	1.50	4.74	1.92	4.81	1.99	170
180	4.54	1.67	5.02	2.14	5.09	2.21	180
190	4.79	1.84	5.30	2.36	5.38	2.44	190
200	5.04	2.03	5.58	2.60	5.66	2.68	200
220	5.54	2.42	6.14	3.10	6.22	3.20	220
240	6.05	2.84	6.70	3.64	6.79	3.76	240
260	6.55	3.29	7.26	4.22	7.36	4.36	260
280	7.06	3.78	7.81	4.84	7.92	5.00	280
300	7.56	4.29	8.37	5.50	8.49	5.68	300
320	8.06	4.84	8.93	6.20	9.05	6.40	320
340	8.57	5.41	9.49	6.93	9.62	7.17	340
360	9.07	6.01	10.05	7.70	10.18	7.96	360
380	9.58	6.65	10.60	8.51	10.75	8.80	380
400	10.08	7.31	11.16	9.36	11.32	9.68	400
420	10.59	8.00	11.72	10.25	11.88	10.59	420
440	11.09	8.72	12.28	11.17	12.45	11.54	440
460	11.59	9.46	12.84	12.12	13.01	12.53	460
480	12.10	10.24	13.40	13.12	13.58	13.56	480
500	12.60	11.04	13.95	14.15	14.14	14.62	500
550	13.86	13.17	15.35	16.88	15.56	17.44	550
600	15.12	15.47	16.74	19.82	16.97	20.49	600
650	16.38	17.94	18.14	22.99	18.39	23.76	650
700	17.64	20.57	19.53	26.36	19.80	27.25	700
750	18.90	23.38	20.93	29.95	21.22	30.96	750
800	20.16	26.34	22.33	33.75	22.63	34.89	800
850	21.42	29.47	23.72	37.76	24.05	39.03	850
900	22.68	32.75	25.12	41.97	25.46	43.38	900
950	23.94	36.20	26.51	46.38	26.88	47.95	950
1000	25.20	39.80	27.91	51.00	28.29	52.72	1000
1100	27.72	47.48	30.70	60.83	31.12	62.89	1100

FRICTION OF WATER IN PLASTIC PIPE
6 inch
C = 150

Flow US GPM	Schedule 40 6.065 in. I.D.		Schedule 80 5.761 in. I.D.		Schedule 120 5.501 in. I.D.		Flow US GPM
	Velocity ft/sec	Head loss ft/100 ft	Velocity ft/sec	Head loss ft/100 ft	Velocity ft/sec	Head loss ft/100 ft	
50	0.56	0.02	0.62	0.03	0.67	0.03	50
60	0.67	0.03	0.74	0.04	0.81	0.05	60
70	0.78	0.04	0.86	0.05	0.94	0.06	70
80	0.89	0.05	0.98	0.07	1.08	0.08	80
90	1.00	0.06	1.11	0.08	1.21	0.10	90
100	1.11	0.08	1.23	0.10	1.35	0.12	100
120	1.33	0.11	1.48	0.14	1.62	0.17	120
140	1.55	0.14	1.72	0.18	1.89	0.23	140
160	1.78	0.18	1.97	0.23	2.16	0.29	160
180	2.00	0.23	2.22	0.29	2.43	0.37	180
200	2.22	0.28	2.46	0.35	2.70	0.44	200
220	2.44	0.33	2.71	0.42	2.97	0.53	220
240	2.67	0.39	2.95	0.50	3.24	0.62	240
260	2.89	0.45	3.20	0.58	3.51	0.72	260
280	3.11	0.51	3.45	0.66	3.78	0.83	280
300	3.33	0.58	3.69	0.75	4.05	0.94	300
320	3.55	0.66	3.94	0.85	4.32	1.06	320
340	3.78	0.74	4.18	0.95	4.59	1.18	340
360	4.00	0.82	4.43	1.05	4.86	1.32	360
380	4.22	0.91	4.68	1.16	5.13	1.46	380
400	4.44	1.00	4.92	1.28	5.40	1.60	400
450	5.00	1.24	5.54	1.59	6.07	1.99	450
500	5.55	1.50	6.15	1.93	6.75	2.42	500
550	6.11	1.79	6.77	2.30	7.42	2.88	550
600	6.66	2.11	7.38	2.71	8.10	3.39	600
650	7.22	2.44	8.00	3.14	8.77	3.93	650
700	7.77	2.80	8.62	3.60	9.45	4.51	700
750	8.33	3.18	9.23	4.09	10.12	5.12	750
800	8.88	3.59	9.85	4.61	10.80	5.77	800
850	9.44	4.01	10.46	5.15	11.47	6.45	850
900	9.99	4.46	11.08	5.73	12.15	7.17	900
950	10.55	4.93	11.69	6.33	12.82	7.93	950
1000	11.11	5.42	12.31	6.96	13.50	8.72	1000
1100	12.22	6.47	13.54	8.30	14.85	10.40	1100
1200	13.33	7.60	14.77	9.75	16.20	12.21	1200
1300	14.44	8.81	16.00	11.31	17.55	14.16	1300
1400	15.55	10.10	17.23	12.97	18.90	16.24	1400
1500	16.66	11.48	18.46	14.74	20.25	18.45	1500
1600	17.77	12.93	19.69	16.61	21.60	20.79	1600
1700	18.88	14.47	20.92	18.58	22.95	23.26	1700
1800	19.99	16.08	22.15	20.65	24.30	25.86	1800
1900	21.10	17.77	23.39	22.83	25.65	28.58	1900
2000	22.21	19.54	24.62	25.10	27.00	31.42	2000
2200	24.43	23.31	27.08	29.94	29.70	37.48	2200
2400	26.65	27.38	29.54	35.16	32.40	44.02	2400

NOMINAL PIPE SIZE: 4 INCH
BASED ON 100 FEET OF PIPE

	C = 100 SCH 40 CI 4.026 I.D.		PVC PIPE SDR21 4.072 I.D.		C = 150 SDR26 4.154 I.D.		SDR35 3.97 I.D.	
GPM	VEL.	FRIC.	VEL.	FRIC.	VEL.	FRIC.	VEL.	FRIC.
40	1.01	0.22	0.99	0.10	0.95	0.09	1.04	0.11
60	1.51	0.47	1.48	0.21	1.42	0.19	1.56	0.24
75	1.89	0.71	1.85	0.31	1.78	0.29	1.94	0.36
100	2.52	1.20	2.46	0.54	2.37	0.49	2.59	0.61
125	3.15	1.82	3.08	0.81	2.96	0.74	3.24	0.92
150	3.78	2.55	3.70	1.14	3.55	1.03	3.89	1.29
175	4.41	3.39	4.31	1.51	4.14	1.37	4.54	1.71
200	5.04	4.34	4.93	1.94	4.73	1.76	5.18	2.19
225	5.67	5.39	5.54	2.41	5.33	2.19	5.83	2.73
250	6.30	6.56	6.16	2.93	5.92	2.66	6.48	3.31
275	6.93	7.82	6.78	3.49	6.51	3.17	7.13	3.95
300	7.56	9.19	7.39	4.10	7.10	3.72	7.78	4.64
350	8.82	12.23	8.62	5.46	8.29	4.95	9.07	6.18
400	10.08	15.66	9.85	6.99	9.47	6.35	10.37	7.91
450	11.34	19.47	11.09	8.70	10.65	7.89	11.66	9.84
500	12.60	23.67	12.32	10.57	11.84	9.59	12.96	11.96
550	13.86	28.24	13.55	12.61	13.02	11.44	14.26	14.27
600	15.12	33.18	14.78	14.81	14.20	13.44	15.55	16.76

NOMINAL PIPE SIZE: 6 INCH
BASED ON 100 FEET OF PIPE

	C = 100 SCH 40 CI 6.065 I.D.		PVC PIPE SDR21 5.993 I.D.		C = 150 SDR26 6.115 I.D.		SDR35 5.91 I.D.	
GPM	VEL.	FRIC.	VEL.	FRIC.	VEL.	FRIC.	VEL.	FRIC.
100	1.11	0.16	1.14	0.08	1.09	0.07	1.17	0.09
125	1.39	0.25	1.42	0.12	1.37	0.11	1.46	0.13
150	1.67	0.35	1.71	0.17	1.64	0.16	1.75	0.19
175	1.94	0.46	1.99	0.23	1.91	0.21	2.05	0.25
200	2.22	0.59	2.27	0.30	2.18	0.27	2.34	0.32
225	2.50	0.73	2.56	0.37	2.46	0.33	2.63	0.39
250	2.78	0.89	2.84	0.45	2.73	0.40	2.92	0.48
275	3.05	1.07	3.13	0.53	3.00	0.48	3.22	0.57
300	3.33	1.25	3.41	0.63	3.28	0.57	3.51	0.67
350	3.89	1.67	3.98	0.83	3.82	0.76	4.09	0.89
400	4.44	2.13	4.55	1.07	4.37	0.97	4.68	1.14
450	5.00	2.65	5.12	1.33	4.92	1.20	5.26	1.42
500	5.55	3.22	5.69	1.61	5.46	1.46	5.85	1.73
550	6.11	3.85	6.26	1.92	6.01	1.74	6.43	2.06
600	6.66	4.52	6.82	2.26	6.55	2.05	7.02	2.42
650	7.22	5.24	7.39	2.62	7.10	2.38	7.60	2.80
700	7.77	6.01	7.96	3.01	7.65	2.73	8.19	3.22
750	8.33	6.83	8.53	3.42	8.19	3.10	8.77	3.66
800	8.88	7.70	9.10	3.85	8.74	3.49	9.36	4.12
850	9.44	8.61	9.67	4.31	9.29	3.91	9.94	4.61
900	9.99	9.57	10.24	4.79	9.83	4.34	10.53	5.12
950	10.55	10.58	10.81	5.29	10.38	4.80	11.11	5.66
1000	11.11	11.64	11.37	5.82	10.92	5.28	11.70	6.23
1050	11.66	12.74	11.94	6.37	11.47	5.78	12.28	6.82
1100	12.22	13.88	12.51	6.94	12.02	6.30	12.87	7.43
1150	12.77	15.07	13.08	7.54	12.56	6.84	13.45	8.07
1200	13.33	16.31	13.65	8.16	13.11	7.40	14.03	8.73
1250	13.88	17.59	14.22	8.80	13.66	7.98	14.62	9.42
1300	14.44	18.92	14.79	9.46	14.20	8.58	15.20	10.13
1400	15.55	21.70	15.92	10.85	15.29	9.84	16.37	11.62

NOMINAL PIPE SIZE: 8 INCH
BASED ON 100 FEET OF PIPE

GPM	C = 100 SCH 40 CI 7.981 I.D.		PVC PIPE SDR21 7.805 I.D.		C = 150 SDR26 7.961 I.D.		SDR35 7.92 I.D.	
	VEL.	FRIC.	VEL.	FRIC.	VEL.	FRIC.	VEL.	FRIC.
175	1.12	0.12	1.17	0.06	1.13	0.06	1.14	0.06
200	1.28	0.16	1.34	0.08	1.29	0.07	1.30	0.08
225	1.44	0.19	1.51	0.10	1.45	0.09	1.47	0.09
250	1.60	0.23	1.68	0.12	1.61	0.11	1.63	0.12
275	1.76	0.28	1.84	0.15	1.77	0.13	1.79	0.14
300	1.92	0.33	2.01	0.17	1.93	0.16	1.95	0.16
350	2.24	0.44	2.35	0.23	2.26	0.21	2.28	0.21
400	2.57	0.56	2.68	0.29	2.58	0.27	2.60	0.27
450	2.89	0.70	3.02	0.37	2.90	0.33	2.93	0.34
500	3.21	0.85	3.35	0.45	3.22	0.40	3.26	0.42
550	3.53	1.01	3.69	0.53	3.55	0.48	3.58	0.50
600	3.85	1.19	4.02	0.62	3.87	0.57	3.91	0.58
650	4.17	1.38	4.36	0.72	4.19	0.66	4.23	0.68
700	4.49	1.58	4.69	0.83	4.51	0.76	4.56	0.77
750	4.81	1.80	5.03	0.94	4.83	0.86	4.88	0.88
800	5.13	2.02	5.36	1.06	5.16	0.97	5.21	0.99
850	5.45	2.26	5.70	1.19	5.48	1.08	5.54	1.11
900	5.77	2.52	6.04	1.32	5.80	1.20	5.86	1.23
950	6.09	2.78	6.37	1.46	6.12	1.33	6.19	1.36
1000	6.41	3.06	6.71	1.61	6.45	1.46	6.51	1.50
1050	6.73	3.35	7.04	1.76	6.77	1.60	6.84	1.64
1100	7.05	3.65	7.38	1.92	7.09	1.74	7.16	1.79
1150	7.38	3.96	7.71	2.09	7.41	1.89	7.49	1.94
1200	7.70	4.29	8.05	2.26	7.73	2.05	7.81	2.10
1250	8.02	4.63	8.38	2.43	8.06	2.21	8.14	2.27
1300	8.34	4.97	8.72	2.62	8.38	2.38	8.47	2.44
1400	8.98	5.71	9.39	3.00	9.02	2.73	9.12	2.80
1500	9.62	6.48	10.06	3.41	9.67	3.10	9.77	3.18
1600	10.26	7.31	10.73	3.84	10.31	3.49	10.42	3.58
1800	11.54	9.09	12.07	4.78	11.60	4.34	11.72	4.45
2000	12.83	11.05	13.41	5.81	12.89	5.28	13.02	5.41
2200	14.11	13.18	14.75	6.93	14.18	6.30	14.33	6.46
2400	15.39	15.48	16.09	8.14	15.47	7.40	15.63	7.59

NOMINAL PIPE SIZE: 10 INCH
BASED ON 100 FEET OF PIPE

	C = 100 SCH 40 CI 10.02 I.D.		PVC PIPE SDR21 9.728 I.D.		C = 150 SDR26 9.924 I.D.		SDR35 9.9 I.D.	
GPM	VEL.	FRIC.	VEL.	FRIC.	VEL.	FRIC.	VEL.	FRIC.
250	1.02	0.08	1.08	0.04	1.04	0.04	1.04	0.04
275	1.12	0.09	1.19	0.05	1.14	0.05	1.15	0.05
300	1.22	0.11	1.29	0.06	1.24	0.05	1.25	0.05
350	1.42	0.14	1.51	0.08	1.45	0.07	1.46	0.07
400	1.63	0.19	1.73	0.10	1.66	0.09	1.67	0.09
450	1.83	0.23	1.94	0.13	1.87	0.11	1.88	0.12
500	2.03	0.28	2.16	0.15	2.07	0.14	2.08	0.14
550	2.24	0.33	2.37	0.18	2.28	0.17	2.29	0.17
600	2.44	0.39	2.59	0.21	2.49	0.19	2.50	0.20
650	2.64	0.46	2.81	0.25	2.70	0.23	2.71	0.23
700	2.85	0.52	3.02	0.28	2.90	0.26	2.92	0.26
750	3.05	0.59	3.24	0.32	3.11	0.29	3.13	0.30
800	3.25	0.67	3.45	0.36	3.32	0.33	3.33	0.33
850	3.46	0.75	3.67	0.41	3.53	0.37	3.54	0.37
900	3.66	0.83	3.88	0.45	3.73	0.41	3.75	0.42
950	3.87	0.92	4.10	0.50	3.94	0.45	3.96	0.46
1000	4.07	1.01	4.32	0.55	4.15	0.50	4.17	0.51
1050	4.27	1.11	4.53	0.60	4.36	0.55	4.38	0.55
1100	4.48	1.21	4.75	0.66	4.56	0.60	4.58	0.60
1150	4.68	1.31	4.96	0.71	4.77	0.65	4.79	0.66
1200	4.88	1.42	5.18	0.77	4.98	0.70	5.00	0.71
1250	5.09	1.53	5.40	0.83	5.18	0.76	5.21	0.77
1300	5.29	1.64	5.61	0.90	5.39	0.81	5.42	0.82
1400	5.70	1.89	6.04	1.03	5.81	0.93	5.84	0.94
1500	6.10	2.14	6.47	1.17	6.22	1.06	6.25	1.07
1600	6.51	2.42	6.91	1.32	6.64	1.19	6.67	1.21
1800	7.32	3.00	7.77	1.64	7.47	1.49	7.50	1.50
2000	8.14	3.65	8.63	1.99	8.30	1.81	8.34	1.83
2200	8.95	4.36	9.50	2.37	9.13	2.15	9.17	2.18
2400	9.76	5.12	10.36	2.79	9.95	2.53	10.00	2.56
2600	10.58	5.94	11.22	3.23	10.78	2.94	10.84	2.97
2800	11.39	6.81	12.09	3.71	11.61	3.37	11.67	3.41
3000	12.21	7.74	12.95	4.22	12.44	3.83	12.50	3.87
3200	13.02	8.72	13.81	4.75	13.27	4.31	13.34	4.36
3400	13.83	9.76	14.68	5.32	14.10	4.82	14.17	4.88
3600	14.65	10.85	15.54	5.91	14.93	5.36	15.00	5.43
3800	15.46	11.99	16.40	6.53	15.76	5.93	15.84	6.00

NOMINAL PIPE SIZE: 12 INCH
BASED ON 100 FEET OF PIPE

	C = 100 SCH 40 CI 11.938 I.D.		PVC PIPE SDR21 11.54 I.D.		C = 150 SDR26 11.77 I.D.		SDR35 11.78 I.D.	
GPM	VEL.	FRIC.	VEL.	FRIC.	VEL.	FRIC.	VEL.	FRIC.
350	1.00	0.06	1.07	0.03	1.03	0.03	1.03	0.03
400	1.15	0.08	1.23	0.04	1.18	0.04	1.18	0.04
450	1.29	0.10	1.38	0.05	1.33	0.05	1.32	0.05
500	1.43	0.12	1.53	0.07	1.47	0.06	1.47	0.06
550	1.58	0.14	1.69	0.08	1.62	0.07	1.62	0.07
600	1.72	0.17	1.84	0.09	1.77	0.08	1.77	0.08
650	1.86	0.19	1.99	0.11	1.92	0.10	1.91	0.10
700	2.01	0.22	2.15	0.12	2.06	0.11	2.06	0.11
750	2.15	0.25	2.30	0.14	2.21	0.13	2.21	0.13
800	2.29	0.29	2.45	0.16	2.36	0.14	2.36	0.14
850	2.44	0.32	2.61	0.18	2.51	0.16	2.50	0.16
900	2.58	0.35	2.76	0.20	2.65	0.18	2.65	0.18
950	2.72	0.39	2.91	0.22	2.80	0.20	2.80	0.20
1000	2.87	0.43	3.07	0.24	2.95	0.22	2.94	0.22
1050	3.01	0.47	3.22	0.26	3.10	0.24	3.09	0.24
1100	3.15	0.51	3.37	0.29	3.24	0.26	3.24	0.26
1150	3.30	0.56	3.53	0.31	3.39	0.28	3.39	0.28
1200	3.44	0.60	3.68	0.34	3.54	0.31	3.53	0.30
1250	3.58	0.65	3.83	0.36	3.69	0.33	3.68	0.33
1300	3.73	0.70	3.99	0.39	3.83	0.35	3.83	0.35
1400	4.01	0.80	4.29	0.45	4.13	0.41	4.12	0.41
1500	4.30	0.91	4.60	0.51	4.42	0.46	4.42	0.46
1600	4.59	1.03	4.91	0.57	4.72	0.52	4.71	0.52
1800	5.16	1.28	5.52	0.71	5.31	0.65	5.30	0.65
2000	5.73	1.56	6.13	0.87	5.90	0.79	5.89	0.78
2200	6.31	1.86	6.75	1.03	6.49	0.94	6.48	0.94
2400	6.88	2.18	7.36	1.21	7.08	1.10	7.07	1.10
2600	7.45	2.53	7.98	1.41	7.67	1.28	7.65	1.27
2800	8.03	2.90	8.59	1.62	8.26	1.47	8.24	1.46
3000	8.60	3.30	9.20	1.84	8.85	1.67	8.83	1.66
3200	9.17	3.72	9.82	2.07	9.44	1.88	9.42	1.87
3400	9.75	4.16	10.43	2.32	10.03	2.10	10.01	2.10
3600	10.32	4.63	11.04	2.57	10.62	2.34	10.60	2.33
3800	10.89	5.11	11.66	2.85	11.21	2.59	11.19	2.57
4000	11.47	5.62	12.27	3.13	11.80	2.84	11.78	2.83
4500	12.90	6.99	13.80	3.89	13.27	3.54	13.25	3.52
5000	14.33	8.50	15.34	4.73	14.74	4.30	14.72	4.28
5500	15.76	10.14	16.87	5.64	16.22	5.13	16.19	5.11

MANUFACTURERS

MANUFACTURING SOURCES FOR SUBMERSIBLE SEWAGE PUMPING SYSTEMS

ABS PUMPS, INC.
140 Pond View Drive
Meriden, CT 06450
Phone: 203/238-2700

ANI PUMPS, INC.
5730 Oakbrook Parkway
Suite 150
Norcross, GA 30093
Phone: 404/441-2425

DAVIS-EMU
1828 Metcalf Avenue
Post Office Box 1419
Thomasville, GA 31792
Phone: 912/226-5733

ENPO PUMP COMPANY
420 East Third Street
Piqua, OH 45356
Phone: 513/773-2442

ENVIRONMENT/ONE CORPORATION
2773 Balltown Road
Post Office Box 773
Schenectady, NY 12301
Phone: 518/346-6161

FAIRBANKS MORSE PUMP CORPORATION
3601 Fairbanks Avenue
Kansas City, KS 66110
Phone: 913/371-5000

FLYGT CORPORATION
129 Glover Avenue
Norwalk, CT 06856
Phone: 203/846-2051

GOULDS PUMPS, INC.
240 Fall Street
Seneca Falls, NY 13148
Phone: 315/568-2811

KSB, INC.
175 Commerce Drive
Hauppage, NY 11788
Phone: 516/231-0303

KOMLINE-SANDERSON ENGINEERING CORPORATION
Post Office Box 257
12 Holland Avenue
Peapack, NJ 07977
Phone: 201/234-1000

THE MARLEY PUMP COMPANY HYDROMATIC PUMPS
5800 Foxridge Drive
Mission, KS 66202
Phone: 913/722-1485

F. E. MYERS COMPANY
Division of McNeil Corporation
400 Orange Street
Ashland, OH 44805
Phone: 419/289-1144

PEABODY BARNES, INC.
651 North Main Street
Mansfield, OH 44902
Phone: 419/522-1511

PUMPTRON DIVISION TRANSAMERICA DELAVAL
829 Bancroft Way
Post Office Box 2007
Berkeley, CA 94170
Phone: 415/843-9400

YEOMANS CHICAGO CORP.
1999 North Ruby Street
Melrose Park, IL 60160
Phone: 312/344-9600

MANUFACTURERS OF COMPONENTS FOR SUBMERSIBLE SEWAGE PUMPING SYSTEMS

THE BILCO COMPANY
Post Office Box 1203
New Haven, CT 06505
Phone: 203/934-6363
(Access Covers)

ELECTRIC SPECIALTY, INC.
Post Office Box 13203A
Orlando, FL 32859
Phone: 305/855-9486
(Controls & Covers)

FLOMATIC CORPORATION
North Hoosick, NY 12133
Phone: 518/686-7381
(Valves)

GENERAL ELECTRIC COMPANY
2000 Taylor Street
Post Office Box 2205
Fort Wayne, IN 46801
Phone: 219/428-4641
(Motors)

PAC-SEAL, INC.
211 Frontage Road
Burr Ridge, IL 60521
Phone: 312/986-0430
(Seals)

PUTNAM WATER GUARD, INC.
Sodom Road
Brewster, NY 10509
Phone: 914/279-5026
(Controls)

RELIANCE ELECTRIC COMPANY
24701 Euclid Avenue
Cleveland, OH 44117
Phone: 216/266-7000
(Motors)

S. J. ELECTRO SYSTEMS, INC.
Route 1, Box 17
Detroit Lakes, MN 56501
Phone: 218/847-1317
(Controls)

STA-CON, INC.
2525 South Orange Blossom Trail
Apopka, FL 32703
Phone: 305/298-5940
(Controls)

INDEX

A

Access covers, 9, 66, 68
Across line starters, 50
Axial flow, 27

B

Ball valve, 63
Ball check valve, 62
Brake horsepower, 26
Bubbler systems, 54
Built-in-place stations, 8

C

Cavitation, 31
Centrifugal pumps, 27
Check valves, 63
Cutwater, 28

D

Design periods, 13
Diaphragm type switches, 53
Discharge size, 6
Duplex stations, 6, 37, 47

E

Electrode controls, 55
Enclosed non-clog impeller, 28

F

Flow
   Average daily, 14
   Maximum daily, 14
   Minimum daily, 14
   Peak hourly, 14
Francis type impeller, 27
Friction head, 34
Friction losses, 30

G

Gate valves, 65
Grinder pumps, 6, 16, 60
Ground fault interruptors, 51
Guide rail systems, 8, 68

H

Hazardous areas, 56

I

Impeller types, 28

L

Large lift stations, 6, 21
Leakage losses, 31
Level control systems, 8, 52
Lift station types, 16
Liquid level controls, 8, 52

M

Mechanical losses, 31
Mechanical shaft seals, 33
Medium lift stations, 5, 17
Mercury float switches, 52
Mixed flow impeller, 27
Motor and motor cavity, 13
Motor controllers, 50
Motors, 7

N

NEMA enclosures, 59
Net positive suction head, 22, 31

O

Operation sequences, 57

P

Plug valves, 64
Population growth projections, 13
Power failure alarms, 58
Prefabricated stations, 8
Pressure sewers, 16
Pressure transducers, 55
Primary power sources, 49
Propeller-type impeller, 27
Pump affinity laws, 26, 30
Pump cycle volumes, 20
Pump failure alarms, 58
Pump rotation, 69
Pump storage, 67
Pump testing, 71

Q

Quick-disconnect sealing flange, 9, 68

R

Radial type impeller, 27
Reduced voltage starters, 51
Redundant Off level with alarm, 58
Residential lift stations, 16

S

Semi-open non-clog impeller, 28
Septic Tank Effluent Pump (STEP), 16
Shock losses, 31
Shut-off valves, 64
Simulation testing, pumps, 74
Siphoning, 32
Site considerations, 10
Small lift stations, 5
Specific speed, 27
Standby power sources, 49
Strainers, 65
Submergence, 32
Swing disc valves, 63

T

Telemetry systems, 58
Trash baskets, 65
Theoretical head, 25
Total dynamic head, 34

U

Ultrasonic level detectors, 55

V

Variable speed controllers, 52
Volute pump, 28
Vortex (recessed) impeller, 28
Vortexing, 32

W

Wastewater flow, 14
Water hammer, 65
Wiring cable entry, 34